Second National Forum on Contemporary Space Research (NaFSKI 2021)

Journal of Physics: Conference Series Volume 2255

Online
7 – 9 October 2021

ISBN: 978-1-7138-5509-5
ISSN: 1742-6588

Printed from e-media with permission by:

Curran Associates, Inc.
57 Morehouse Lane
Red Hook, NY 12571

proceedings.com

Some format issues inherent in the e-media version may also appear in this print version.

Institute of Physics
Dirac House, Temple Back
Bristol BS1 6BE UK

Phone: 44 1 17 929 7481
Fax: 44 1 17 920 0979

techtracking@iop.org

Additional copies of this publication are available from:

Curran Associates, Inc.
57 Morehouse Lane
Red Hook, NY 12571 USA
Phone: 845-758-0400
Fax: 845-758-2633
Email: curran@proceedings.com
Web: www.proceedings.com

TABLE OF CONTENTS

2nd National Forum on Contemporary Space Research (NaFSKI 2021), October 7-9, 2021, Sofia TechPark, Bulgaria

Peer Review Statement

Why Total Solar Eclipses Are Important to Science? 1
Ts Tsvetkov, N Petrov

Observational Features of Thin Accretion Disks Around Traversable Wormholes 8
V Deliyski, G Gyulchev, P Nedkova, S Yazadjiev

Galactic Cosmic Rays Above the Earth's Atmosphere 14
Marusya Buchvarova

Model of Galactic Cosmic Ray Spectrum Above the Earth's Atmosphere 29
M Buchvarova, D Draganov

Observational Features of Hot Spots on Neutron Stars with a Photon Sphere 34
Galin Gyulchev

Coupled-Cluster Calculation of Neutron Matter Equation of State 42
Stoyan Mishev

Some Warnings About Quantum Space Gravimetry Enhance Earth Observations Project 46
Plamen Fiziev

Ofine Noise Calibration of the CAEN V1742 ADCs at the PADME Experiment 58
Georgi Georgiev

PADME Physics Program 64
Radoslav Simeonov

Feasibility Study of Simultaneous Measurement of Multiple Parameters from a Single Monolithic Scintillator Crystal Using the Geant4 Simulation Toolkit 70
S Ivanov, Zh Toneva

Impact Point Determination of Gamma Interaction in a Monolithic Scintillator Crystal Read Out by a SiPM Matrix 75
Svetoslav Ivanov, Simeon Ivanov

Extra-Terrestrial Influence on Climate Variability 79
Ts P Velichkova, N A Kilifarska

Geomagnetic Control of Cosmic Radiation and Its Impact on the Irregularly Distributed Lower Stratospheric Ozone 88
N A Kilifarska, K Peqini

Future Trends of Commercial In- Orbit Satellite Servicing, Active Debris Removal and End-Of Life Services 100
Stella Alexandrova

Analysis of Hyperspectral and Multispectral Reflectance Spectra in the Black Sea Coastal Area Near the Danube Delta: Comparison of PRISMA and Sentinel-2 Observations 108
Irina Gancheva

Author Index

Journal of Physics: Conference Series **2255** (2022) 011001 doi:10.1088/1742-6596/2255/1/011001

2nd National Forum on Contemporary Space Research (NaFSKI 2021), October 7-9, 2021, Sofia TechPark, Bulgaria

The National Forum on Contemporary Space Research is an initiative of the "Cosmos" Branch of the Union of Physicists in Bulgaria. The scientists from several scientific institutions working in the space research field gathered and identified the main problems facing the community in the country: insufficient awareness of the scientific community about the activities carried out in Bulgaria related to space research; poor communication with the general public; lack of science-business connection; underdeveloped training related to the space research conducted in the country, which in turn leads to a shortage of staff in the sector. To address these problems, the "Cosmos" Branch started several initiatives directed to different target groups to promote research networking both on national and international levels, to facilitate discussions with the business in Bulgaria and to connect research groups with state administration, to enhance competition, and create a healthy environment that would boost the Space sector development in Bulgaria. The ultimate goal of "Cosmos" branch is to create a vibrant society of researchers, entrepreneurs, teachers, administration, and everyone who is interested in having Bulgaria into the international space elite. Thus, the National Forum on Contemporary Space Research was shaped as a platforms where the Bulgarian researchers meet, discuss, and collaborate.

The first edition of the Forum took place in 2020 and it proved the idea is timely and successful. That is why the second edition of the Forum was organized as a 3 days event in Sofia TechPark "Incubator" Hall in the period 7-9 October 2021. There was also a possibility to participate on-line due to the COVID-19 restriction on the number of on-site participants.

The event was held under the auspices of the Bulgarian Ministry of Economy with the financial support of the Sofia University St. Kliment Ohridski, "America for Bulgaria" Foundation and "Evrika" Foundation.

The Forum programme was organized in 4 sections: "Business and Cosmos", "Science and Cosmos", "Education and Cosmos" and "Young scientists and PhD students' session". The Forum was opened with a greeting address from the Minister of Economy Mrs. Daniela Vezieva. Representatives of European Space Agency and European Union Agency for the Space Programme also took part in the "Business and Cosmos" session. More than 100 participants from various sectors related to the Space activities participated in the 4 sessions.

The program and presentations during the Forum are available at https://bulgarianspace.online/nafski2021/

The presenters in the "Science and Cosmos" and "Young Scientists and PhD students" sessions were invited to contribute to the Proceedings of the 2nd edition of the National Forum on Contemporary Space Research. The collected papers passed a peer-review process and we are proud to present the collection to the readers of the Journal of Physics – Conference Series. The papers are separated in three categories, according to the topic: fundamental space science research, aerospace technologies and Earth observations.

We are thankful to all participants, sponsors, and guests, who contributed to the Forum and helped us to fulfil one of the most recognized events in this scientific topic this year - without you this would never have happened.

We are looking forward to welcome you again at the 3rd edition in 2022.

Assoc. Prof. Dr. Elisaveta Peneva (acting as a Guest Editor)

Journal of Physics: Conference Series **2255** (2022) 011002 doi:10.1088/1742-6596/2255/1/011002

Peer Review Statement

All papers published in this volume have been reviewed through processes administered by the Editors. Reviews were conducted by expert referees to the professional and scientific standards expected of a proceedings journal published by IOP Publishing.

- **Type of peer review:** Single Anonymous
- **Conference submission management system:** Morressier
- **Number of submissions received:** 16
- **Number of submissions sent for review:** 15
- **Number of submissions accepted:** 15
- **Acceptance Rate (Submissions Accepted / Submissions Received × 100):** 93.8
- **Average number of reviews per paper:** 1
- **Total number of reviewers involved:** 17
- **Contact person for queries:**
  Name: Elisaveta Peneva
  Email: elfa@phys.uni-sofia.bg
  Affiliation: Sofia University St. Kliment Ohridski

Journal of Physics: Conference Series **2255** (2022) 012001 doi:10.1088/1742-6596/2255/1/012001

Why total solar eclipses are important to science?

Ts Tsvetkov and N Petrov

Institute of Astronomy and National Astronomical Observatory, Bulgarian Academy of Sciences

E-mail: tstsvetkov@astro.bas.bg

Abstract. Despite today we have plenty of opportunities for high-quality observations of the Sun from both Earth and space, total solar eclipses remain a rare chance to explore the corona, its features and various atmospheric phenomena. Members of our team took part of 5 out of 6 Bulgarian expeditions for total solar eclipse observations ever realized. In the current paper we summarize the tasks we fulfilled and the scientific results we obtained. Special attention is paid to the investigation of the shadow bands that accompanied the last eclipse we witnessed – on 2019 July 2.

1. Introduction

Solar eclipses impressed people since ancient times and made them look for a way to explain the world they live in. Increasing their knowledge humanity reached a stage of detailed studies of both the Sun and the eclipses. Today, for scientists, they are not just a beautiful sight, but also an incredible chance to conduct important data to reveal the nature of solar corona. It is a widespread opinion that total solar eclipses (TSEs) are a "celestial laboratory" that allows testing of various hypotheses and statements, which otherwise can not be checked on Earth.

The first scientific task ever solved during solar eclipse was the one for the apparent motions of the Sun and the Moon in the sky. Then astronomers focused attention on the Sun and phenomena visible during the totality – solar corona and its changes, chromosphere, prominences, etc. Modern instruments allow accurate measurements of eclipse contacts. Comparing them with the theoretically predicted values allows estimation of additions to the theory of the Moon's motion, its parameters, the Earth's rotation around the axis, etc.

The data from TSE observations might have huge impact not only for different branches of physics and astronomy, but also even for history. Despite the information about past eclipses may be inaccurate and scarce, it often turns out useful for scientists. Many eclipses were described in historical sources because they were interpreted as important signs for the denouement of the events. Thus, they play an important role in dating certain historical episodes like the battle between the Lydians and the Medes in the VI century BC (also called "the battle of the eclipse"). It was described by Herodotus, who mentions the solar eclipse that interrupted the battle and that helped to finally determine the moment in which it happened [1].

Data from past eclipses also provide information on how different nations of antiquity counted time – how many hours they thought there was in a day and how long they considered to last an hour. Using the day and time of a specific eclipse from a chronicle and calculating when (according to our timekeeping) it happened, we can compare our time measuring method with others and use it to date other historical events.

TSEs are the only possibility to observe and explore solar corona in its entirety. Scientists analyse its small-scale structure and watch out for fast changes in different structures (fibers, plasmoids, polar plumes, streamers, etc.). White-light observations help the study of coronal brightness distribution according to heliographic coordinates, distance to the solar limb, the shape of the corona and its flatness [2, 3]. Polarized white-light data can help in determination of local magnetic fields and their direction, and the degree of polarization as a function of electron density in the corona, while spectral analyses reveal information for the state of coronal material and its temperature.

Every TSE is accompanied by changes in the conditions in Earth's atmosphere due to the decrease of the amount of energy reaching our planet. All meteorological parameters (temperature, pressure, humidity, wind speed and direction, clouds formation, etc.) vary during the totality and reveal atmosphere's response.

Six expeditions for observation of TSEs have been fulfilled by Bulgarian scientific teams to this day (Figure 1) – 2021 marks the 60-th anniversary of the first Bulgarian experiments carried out during an eclipse. Another one was planned for December 2020 [4], but it was cancelled because of the international travel restrictions. Members of our team took part in 5 of the expeditions (except for the one in 1961). Using part of the obtained observational data, in the current study we present our point of view on why TSEs are still important to science today.

2. Scientific experiments

Today we have advanced instruments and obtain high-resolution ground- and space-based observations, but TSEs remain invaluable for solving various specific tasks.

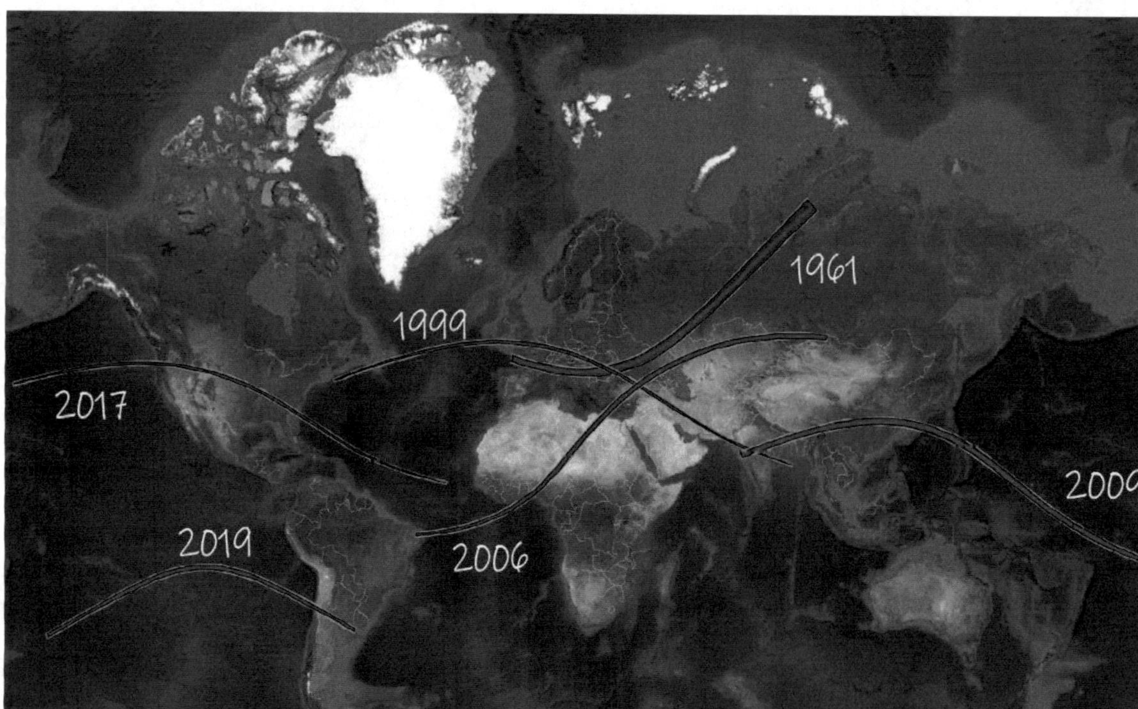

Figure 1. Map of totality paths of all TSEs observed by Bulgarian scienftific teams. Solar eclipses paths across the Earth's surface are obtained by Xavier Jubier (http://xjubier.free.fr)

2.1. Coronal structure

The intensity of the solar corona (many times less than that of the photosphere) makes it difficult to observe. The absence of a bright background makes coronal observations at higher distances possible during TSEs. Despite such observations are insecure because of the short duration of the totality, the width of the totality path, the dependence on the meteorological conditions at the time of observation, etc., they remain the most appropriate way to study the corona both in detail and as a whole (Figure 2).

Thanks to the progress of observing equipment and methods, we now know about the existence of a variety of small-scale solar structures. TSE observations with a resolution in order of 1 arc sec demonstrate that the majority of the corona is in a dynamic state [5].

The structure of the corona depends on the heliographic latitude and longitude, the distance to the limb and the phase of solar activity cycle – during the transition from minimum to maximum, its density at the equator increases about 2 times, and the temperature by about 20% [6]. The typical corona observed near solar minimum shows short elongated rays along the equator, while in the epoch of maximum its shape becomes asymmetrical, influenced by strong magnetic fields. A "quiescent" corona cannot be observed even at the poles in an era of minimum. Still open remains the question about another parameter of the corona that can be determined by TSE observations – its flattening. Summarized results from our estimations on coronal flattening performed during four TSEs – 1999 August 11 (Bulgaria), 2006 March 29

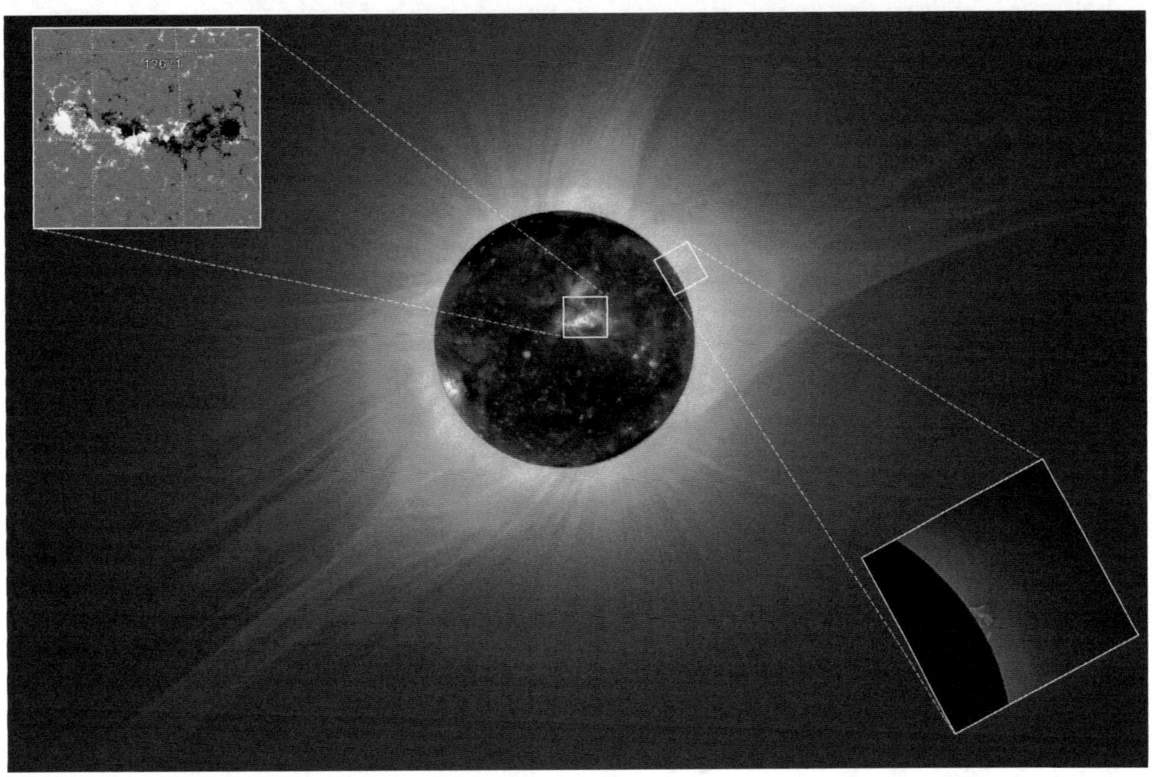

Figure 2. Solar corona from the TSE on 2017 August 21. The image is obtained after combination of plurality of frames with different exposure times, AIA/SDO image of the solar disk in the moment of total eclipse, HMI/SDO magnetogram of the active region and prominence close-up taken with a longer focal length lens.

(Turkey), 2017 August 21 (USA) and 2019 July 2 (Chile) are published in [7].

2.2. Polarization of white-light corona

Polarimetric measurements conducted during TSEs are an important source of information about the physical conditions in the so called white-light corona. During every TSE such measurements are a fundamental experiment in view of the fact that numerous problems related to polarization of the corona are still unsolved [8, 9]. One of those problems is the existence of areas of polarization in the inner corona which exceed the maximum possible one, when considering only the Thomson scattering of photosphere photons. Also there are no unambiguous data on the dependence between the degree of polarization and the wavelength or on its value in the polar plumes.

The investigation of the global characteristics of the solar corona is based on the fact that its white-light intensity is linearly related to the density of free electrons. Thus, photometric observations during TSEs offer possibility for direct determination of electron density in the corona and present the large-scale distribution of the electron concentration.

White-light observations using linear polarizers with possible turning on fixed positions also help the division of the K and F corona components. In distances smaller than 3 solar radii the F-corona (called dust corona as well) is not practically polarized.

Such observations also show the impact of the sky polarization on the brightness of the polarized solar corona, depending on the landscape during the totality and that the singular polarization points of the corona are in linear dependence with the height of the Sun above the horizon. This and more results from our polarized corona experiments are published in [10].

2.3. Active processes and their environment

TSEs are still a significant opportunity for detailed explorations of solar dynamic events taking place in the upper atmosphere and the links between them because of the ability they provide to capture the different parts of the corona simultaneously. The main solar activity events to observe in the lower corona are prominences. They are usually located in the base of coronal streamers, which consist of relatively dark "empty spaces" inside the concentric loop-like structures (Figure 2). The coronal cavities are positioned directly over the quiescent prominences. They are unstructured and inhomogeneous and form about 40% of solar eruptions [11]. Fine, dark fibers are sometimes registered there which can be interpreted as proof of comparatively low-temperature plasma. Monitoring the fine structure, dynamics and destabilization of quiescent prominences helps to register, in the earliest possible stage, the physical conditions and processes leading to the development of eruptive instability and filament eruptions. The eruptive prominences represent a temporary or a final phase of the life of the quiescent ones and are often associated with coronal mass ejections (CMEs) at higher altitudes [12]. Despite our expedition for the 2020 December TSE was cancelled, other observers showed that it offered a rare chance to observe a CME during the totality [13]. As both CMEs and eruptive prominences are among the engines of the space weather, these observations may turn out to be especially important since no ground- or space-based coronagraphs can effectively cover the distance range of 1.3 to 2.2 solar radii, but 74% of CMEs undergo most dramatic acceleration below 1.5 solar radii [14]. These powerful active solar processes are directly related to data alternation of the parameters of the terrestrial magnetosphere and ionosphere.

Our observations and analyses on active solar processes detected during TSEs can be found in [15, 16, 17].

2.4. Shadow bands and related atmospheric conditions

A yet not completely studied atmospheric formations observed a few seconds before and after the totality of every TSE are the shadow bands. Despite they were first mentioned 200 years ago, its successful photo- and videoregistration became possible only in the last few decades. The widespread opinion claims that they are formed when the light from the tiny solar crescent is refracted by turbulent patches in Earth's atmosphere, but the role of different layers of the atmosphere in the pattern of the shadow bands is still unclear.

We successfully registered the shadow bands on 3 of our expeditions – in 2006, 2017 and 2019. Our equipment in the last two cases allowed us to realize a relatively new experiment by not only videotaping the bands, but also monitoring the near-ground atmospheric conditions in order to estimate their impact on the shadow bands pattern. In 2017 our experimental setup for videoregistration included white screen for visualisation of the bands and Sony DCR-SR55 digital video camera (25 frames/second), which was replaced 2 years later by higher-quality camera Sony Alpha 7III (50 frames/second). In both cases we used Gill Windsonic anemometer situated at 1.5 m above the ground that records the wind velocity and direction 4 times/second, but in 2019 we widened the meteorological data we collected by adding temperature and humidity data logger LOG210 to our equipment.

More analyses on the bands registered in 2006 and 2017 is available in [7]. Here we present our results from the experiment performed in 2019, using in addition also the details only for the bands observed before the totality in 2017, because the insufficient quality of the videotaping after the totality does not allow data processing.

Atmospheric conditions we measured in 2019 include temperature, humidity and wind velocity (Figure 3). It is well known that these are parameters with a wide range of variation for every TSE. When considering the weather elements, it is important to note that they are strongly dependent on the peculiarities of the location. Our team was situated in Chile (29°47'37.0"S 70°53'12.4"W) at altitude more than 1400 m.

The lowest temperature (13.0°C) was registered exactly 27 minutes after the end of the totality. It marks temperature drop of 5.4°in comparison with the highest value measured exactly 27 minutes before the second contact. It is important to mention that the altitude of the Sun at moment of maximum eclipse on 2019 July 2 for our location was 13.2°and our observations ended with the sunset (about an hour after the totality). The average temperature

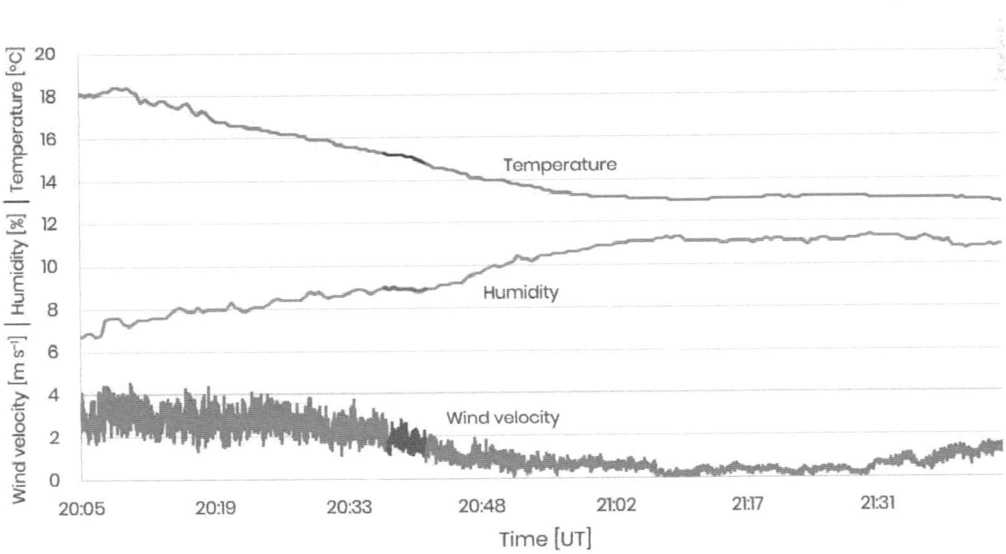

Figure 3. Meteorological parameters during the TSE in 2019. In different colour are presented the values during the totality.

during the total phase of the eclipse was 15.2°.

The humidity is a less explored metorological factor during TSEs than temperature. Our measurements show amplitude of 4.2% as the maximum of the relative humidity is registered about 13 minutes after the third contact.

The average wind velocity shows rapid changes before and after the totality – from 2.27 to 1.59 m s^{-1}, respectively. Nevertheless, the wind direction remains similar (eastern-northeastern). The observed drop in the wind velocity coincides with reduction of the frequency of the observed shadow bands (Figure 4).

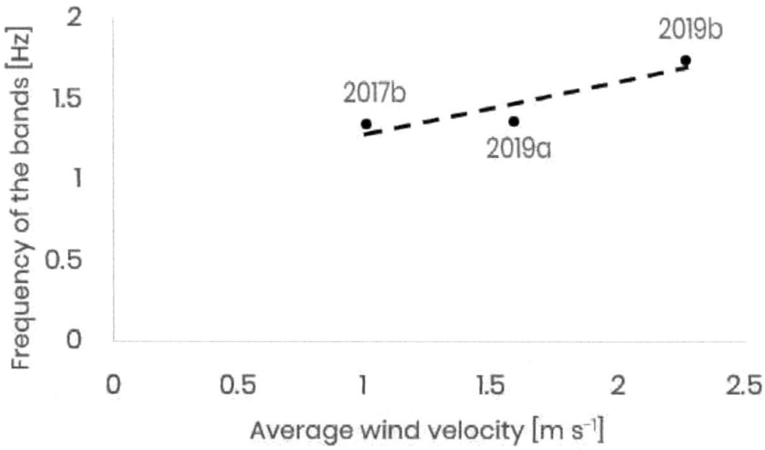

Figure 4. The dependence of shadow bands frequency on average wind speed in the moment of observation. The labels next to each point note the year of the TSE when they were measured at, the letters ("a" and "b") show if the values are obtained after or before the totality of the eclipse, respectively.

3. Conclusions

Total solar eclipses are not only an impressive celestial attraction, but also a real challenge for scientists. There are still a number of open questions related to the solar corona, which can only be studied in detail during total solar eclipses. Precise observations show some already established global parameters of the corona, such as its shape, electron density, physical parameters as temperature and magnetic field, as well as the presence of emission lines of highly ionized chemical elements. But modern methods of observation show that the physics of dynamic processes related to the energy transfer in the corona as well as the relationship of local active processes in the solar atmosphere, where powerful eruptions and coronal mass ejections take place, has not yet been well studied. Every total solar eclipse is unique and no complete recurrence is ever observed. The phase of the 11-year solar cycle is essential for the expected local active processes on the Sun and in the corona. The presence of these events is the basis of solar-terrestrial influences and the determination of the space weather.

At the same time, the study of solar corona is an important part of astronomy due to the fact that the models for its formation and dynamics are the basis for studying and modelling of all other stellar atmospheres of all stars in our and other galaxies. This is our most powerful stellar laboratory, which we can study in detail better than any other star.

Finally, the occulted Sun provides an opportunity to witness some unique accompanying events and changes in atmospheric conditions on Earth unrepresentative for the moments beside total eclipses. Shadow bands are known for centuries, but even today we are not sure about the role of the near-surface weather conditions on their pattern. Our latest experiment shows coincidence between the wind velocity and frequency of the observed bands. Nevertheless, the question remains unsolved until future eclipses reveal additional data.

Acknowledgments
The study is supported by NSF with contract No. KP-06-H28/4 and Bulgarian Ministry of Education and Science under the National Research Programme "Young scientists and postdoctoral students" approved by DCM #577/17.08.2018.

References
[1] Panchenko D 1994 *J. Hist. Astron.* 275.
[2] Gulyaev R A 1997 *AApTr* **13(2)** 137.
[3] Ludendorff H 1928 *Phys.-Math. IClasse* **16** 185.
[4] Tsvetkov Ts, Petrov N, Miteva R, Ivanov E and Popov V 2020 *Proc. of the 12th Workshop "Solar Influences on the Magnetosphere, Ionosphere and Atmosphere"* (Primorsko) p 40.
[5] Pasachoff J, Rusin V, Druckmuller M and Saniga M 2007 *ApJ* **665(1)** 824.
[6] Priest E 2014 *Magnetohydrodynamics of the Sun* (St Andrews: University of St Andrews).
[7] Tsvetkov Ts, Miteva R, Ivanov E, Popov V, Nakeva Y, Bojevski L, Damm T and Petrov N 2019 *Proc. of SES 2019* (Sofia) 52.
[8] Badalyan O G, Livshits M A and Sykora J 1997 *Sol. Phys.* **173(1)** 67.
[9] Molodensky M M, Starkova L I, Kutvitskii V A and Merzlyakov V L 2009 *Astron. Rep.* **53(3)** 275.
[10] Merzlyakov V L, Tsvetkov Ts, Starkova L I and Petrov N 2019 *Serb. Astron. J.* **199** 83.
[11] Pevtsov A, Panasenco O and Martin S 2012 *Sol. Phys.* **277(1)**.
[12] Hori K and Culhane J L 2002 *Astron. Astrophys.* **382** 666.
[13] Boe B, Yamashiro B, Druckmuller M and Habbal S 2021 *ApJ* **914(2)** L39.
[14] Bein B M, Berkebile-Stoiser S, Veronig A M, Temmer M, Muhr N, Kienreich I, Utz D and Vršnak B. 2011 *ApJ* **738** 191.
[15] Tsvetkov Ts, Petrov N and Shirov G 2021 *Proc. of the 13th Workshop "Solar Influences on the Magnetosphere, Ionosphere and Atmosphere"* (Primorsko), *in press.*
[16] Tsvetkov Ts, Petrov N, Myshyakov I and Miteva R 2019 *Solar prominence observations during 2017 total solar eclipse* (Sofia: VarSITI Closing Symposium).
[17] Tsvetkov Ts 2020 *Bulg. Astron. J.* **33** 117.

Observational features of thin accretion disks around traversable wormholes

V Deliyski[1,a], **G Gyulchev**[1,b], **P Nedkova**[1,c] **and S Yazadjiev**[1,2,d]

[1]Department of Theoretical Physics, Faculty of Physics, Sofia University, Sofia 1164, Bulgaria
[2]Institute of Mathematics and Informatics, Bulgarian Academy of Sciences, Acad. G. Bonchev 8, 1113 Sofia, Bulgaria

E-mail: [a] Valentin.Deliyski@gmail.com
[b] gyulchev@phys.uni-sofia.bg
[c] pnedkova@phys.uni-sofia.bg
[d] yazad@phys.uni-sofia.bg

Abstract. We investigate the potential appearance of qualitatively new observational features of accretion disk images around compact objects. More specifically we focus on a particular class of traversable wormholes, possessing a photon sphere. We discover that for certain solution parameters the shape of the images produced possesses distinct features that are not present for Kerr black holes. This provides potential avenues for differentiating between compact objects based on observational data at the scale of the gravitational radius.

1. Introduction

Historically the study of gravitational lensing - i.e. the bending of light as is passes through a gravitational field - has been observed only in the linearised regime in the form of multiple lensed images of supernovas and galaxies [1] or in the hunt for exoplanets [2]. Recent collaborations such as the Event Horizon Telescope have shown that it is feasible to carry out observations of supermassive black holes with near horizon scale resolution [3]. This allows us to directly study the gravitational lensing effect at work in the strong gravity regime, where we expect the peculiarities of different spacetimes to become apparent in the images.

We will consider a thermally emitting thin accretion disk surrounding traversable wormholes. The bending of light emitted from the disk is strong enough as to loop the photons around the wormhole multiple times before reaching the observer. This results in a family of images corresponding to the number n of half-turns the photons complete n. These images contain rich information about the underlying spacetime and we are now beginning to reach the observational resolution necessary to start looking at peculiarities in their features.

The aim of the current work is to find parameter values for certain wormhole solutions which give rise to characteristic features of accretion disk images, that can be used to observationally distinguish them from other compact objects, and thus confirm their existence. We consider static spherically symmetric spacetimes and treat the image formation analytically, following [4, 5] and employ a numerical ray tracing scheme along with an analytic model for thin disks, developed by Page and Thorne [6]. We find that the images produced from the disk on our side of the throat are qualitatively similar to those of Schwarzschild black holes, while the images from the other side of the throat give rise to entirely new observational features. These formations are of a size comparable to that of the shadow itself, so with a moderate resolution increase to the imaging process, they can realistically be observed in future collaborations.

2. Traversable wormhole spacetimes

The general form of a spherically symmetric spacetime, describing traversable wormholes is given by the metric, proposed by Teo [7]

$$ds^2 = -N^2(r)dt^2 + \frac{dr^2}{1 - \frac{b(r)}{r}} + r^2K^2(r)\left[d\theta^2 + \sin^2\theta d\phi^2\right] \tag{1}$$

where the lapse and shape functions N and b, as well as the function K must satisfy certain conditions, which ensure traversability. Namely the spacetime must not possess an event horizon, requiring the lapse function N to be non-zero and finite everywhere, and the shape function must satisfy the so called flare out condition $b(r) - r\partial_r b(r)|_{r=r_0} > 0$, where r_0 is the location of the throat.

In studying the images produced by a metric of the form (1) we will differentiate between two types of parameters: observer dependent - the distance to the observer r_{obs} and his or her inclination angle relative to the normal of the disk plane i, and metric dependent - the position of the unstable photon orbits r_{ph} and the position of the ISCO orbit for massive particles r_{ISCO}. While we are free to specify r_{obs} and i independently from anything else, the metric dependent parameters we determine from the equations of motion for a given spacetime.

2.1. Equations of motion

For a general spherically symmetric metric there exist two conserved quantities, arising from the Killing symmetries, governing particle motion: their energy E and angular momentum about the chosen z axis L_z. Separation of variables yields a third conserved quantity Q, which functions as the total angular momentum at infinity. This leads to equations of motion of the form [8]:

$$\frac{N(r)}{\sqrt{1 - \frac{b(r)}{r}}}\frac{dr}{d\lambda} = \pm_r\sqrt{R(r,m)} \tag{2a}$$

$$r^2K^2(r)\frac{d\theta}{d\lambda} = \pm_\theta\sqrt{\Theta(\theta)} \tag{2b}$$

$$\frac{d\phi}{d\lambda} = \frac{L_z}{r^2K^2(r)\sin^2\theta} \tag{2c}$$

$$\frac{dt}{d\lambda} = \frac{E}{N^2(r)}. \tag{2d}$$

Where we define the effective potentials $R(r, m)$ and $\Theta(\theta)$ as:

$$R(r,m) = -Q\frac{N^2(r)}{r^2K^2(r)} + E^2 - m^2N^2(r) \tag{3a}$$

$$\Theta(\theta) = Q - \frac{L_z^2}{\sin^2\theta}, \tag{3b}$$

where m is the mass of the particle. The concrete form for the metric functions we consider is:

$$N(r) = e^{-\gamma\frac{r_0}{r}}, \quad b(r) = r_0, \quad K(r) = 1 \tag{4}$$

2.2. Metric dependent parameters

For a fixed observer the structure of the family of unstable circular photon orbits determine the general shape of the images, while the position of the ISCO orbit is our inner cutoff for the accretion disk.

The conditions for the existence of unstable circular photon orbits outside the throat are:

$$R(r,0)\big|_{r=r_{ph}} = 0, \qquad \frac{dR(r,0)}{dr}\bigg|_{r=r_{ph}} = 0, \qquad \frac{d^2R(r,0)}{dr^2}\bigg|_{r=r_{ph}} \geq 0, \tag{5}$$

while those for orbits on the throat are given by:

$$1 - \frac{b(r)}{r}\bigg|_{r=r_0} = 0, \quad R(r,0)\big|_{r=r_0} = 0, \quad \frac{dR(r,0)}{dr}\bigg|_{r=r_0} \geq 0. \tag{6}$$

Equations (5) and (6) define a parametric relation, which gives the boundary of the wormhole shadow [8, 9]. They also determine the radial turning point structure as a function of the scaling parameter γ. The conditions for ISCO are similar, with the overall requirement $m \neq 0$:

$$R(r,m)\big|_{r=r_{ISCO}} = 0, \qquad \frac{dR(r,m)}{dr}\bigg|_{r=r_{ISCO}} = 0, \qquad \frac{d^2R(r,m)}{dr^2}\bigg|_{r=r_{ISCO}} = 0. \tag{7}$$

For certain values of the scale parameter γ solutions to (7) may not exist outside the throat, and for such cases the stable circular timelike orbits reach up to the wormhole throat $r = r_0$.

3. Analytical ray tracing

For spherically symmetric spacetimes the motion of particles is in general planar, and the definition of the equator $\theta = \pi/2$ is arbitrary. Therefore we can describe the trajectories of photons using just equations (2a) and (2c) restricted to $\theta = \pi/2$, which after introducing the impact parameter $\xi = L_z/E$, reduce to:

$$\left(\frac{dr}{d\phi}\right)^2 = \frac{1}{1-\frac{b(r)}{r}} \frac{r^4}{\xi^2 N^2(r)} V_{eff}, \quad V_{eff}(r) = 1 - \frac{N^2(r)}{r^2}\xi^2, \tag{8}$$

where the radial turning point structure outside the wormhole is determined entirely by V_{eff}. One can now connect the azimuthal displacement of the photon $\Delta\phi$ to the observer polar coordinates on the screen $\{\sigma, \delta\}$ and the inclination i by [10]:

$$I(\xi, r_s, r_{obs}) = n\pi + \cos^{-1}\left(-\frac{\sin\delta\tan i}{\sqrt{\sin^2\delta\tan^2 i + 1}}\right), \tag{9a}$$

$$\sigma = r\sin^{-1}\left[\sqrt{\frac{\xi^2}{r^2}\left(1 - \frac{b(r)}{r}\right)}\right]\bigg|_{r=r_{obs}} \tag{9b}$$

where $I(\xi, r_s, r_{obs})$ is the solution to (8) in the form $\Delta\phi = \phi(r_s) - \phi(r_{obs})$ and r_s denotes the position of the emitting fluid element, reached by the photon with impact parameter ξ. The parameter n counts the image order. The solution to (9a) for fixed $\{\xi, r_s, r_{obs}, i\}$ give us the polar angle δ corresponding to a point on the image, as seen by the observer and (9b) gives us its radial position on the screen. The first 3 images (corresponding to $n = 0, 1, 2$) are given in Fig. 1 for the case of $\gamma = 1$ and $r_s = 12r_0$, where we consider disks on both sides of the wormhole throat. Note that it is standard to give distance measurements in terms of geometric mass, so we have opted to set the axes in units of ADM mass $M_{ADM} = \gamma r_0$

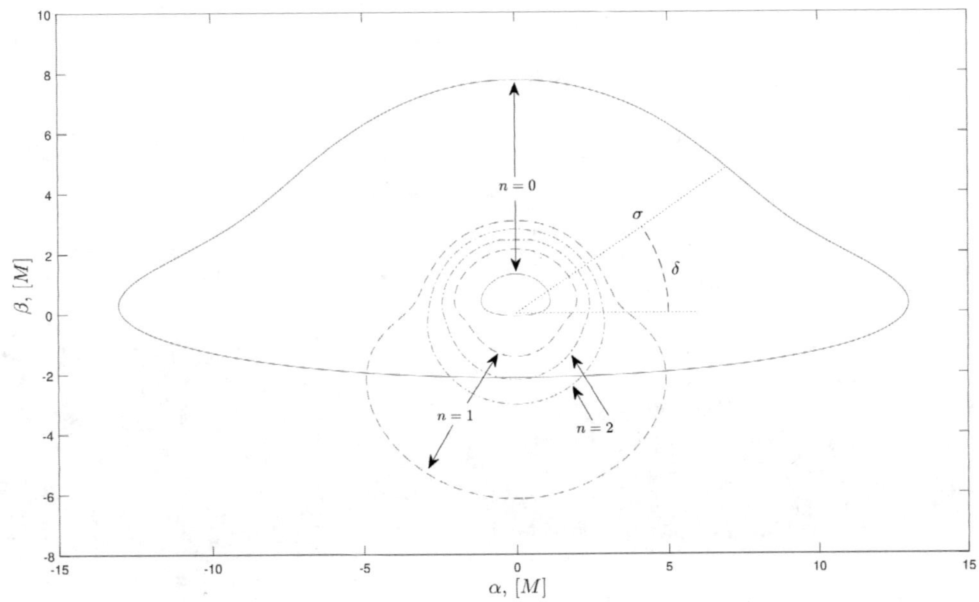

Figure 1. The first three images of a single orbit at $r = 12r_0$ from both sides of the throat (our universe and the other one), as seen by an observer at infinity and inclination $i = 80\,\deg$. The axes are in units of ADM mass.

We find that such an image structure exists for all values of the scaling parameter γ and thus can be considered general for spacetimes of the form (4). Notably other authors have found similar images for different classes of wormholes [11, 12].

4. Numerical ray tracing

Here we present the results of numerically integrating the equations of motion using standard adaptive Runge-Kutta schemes. For numerical stability we consider the dynamics as a Hamiltonian system of equations for the position and momenta k_μ with $H = \frac{1}{2}g^{\mu\nu}k_\mu k_\nu$. The photon trajectories are traced back from the observer, where the initial radial momentum k_r is given as $k_r = g_{rr}\frac{dr}{d\lambda}|_{r=r_{obs}}$ with the help of (2a) and the angular momenta $\{k_\theta, k_\phi = \xi\}$ are varied in a loop to scan a given observational window. When the photon reaches a point on the disk, we project it onto the observational plane using

$$\alpha = -\frac{\xi}{\sin\theta_0}, \tag{10a}$$

$$\beta = \sqrt{\eta - \frac{\xi^2}{\sin^2\theta_0}} \tag{10b}$$

where $\eta = Q/E^2$ is evaluated from the initial momenta $k_\theta|_{r=r_{obs}}$ using (3b) and (2b). We also calculate the radiation flux from the disk F at the emission point, using the model developed by Thorne [6]:

$$F(r) = -\frac{\dot{M}}{4\pi\sqrt{-g}}\frac{\partial_r\Omega_m}{(E-\Omega L_{z,m})^2}\int_{r_{ISCO}}^{r}(E_m - \Omega_m L_{z,m})\partial_r L_{z,m}dr, \tag{11}$$

where Ω_m, $L_{z,m}$ and E_m are the angular velocity and momentum with respect to the rotation axis and energy of the orbiting matter. $\dot{M}$ is the mass accretion rate and g is the determinant of the metric tensor. The observed flux is redshifted by a factor of $\left(\frac{1}{1+z}\right)^4$, given by the expression:

$$\frac{1}{1+z} = \frac{u^{\mu}_{obs} k_{\mu}|_{obs}}{u^{\nu}_{e} k_{\nu}|_{emitter}} \tag{12}$$

where u_e is the four-velocity of the emitter and u_{obs} is that of the observer. Some results from the simulations are shown on Fig. 2.

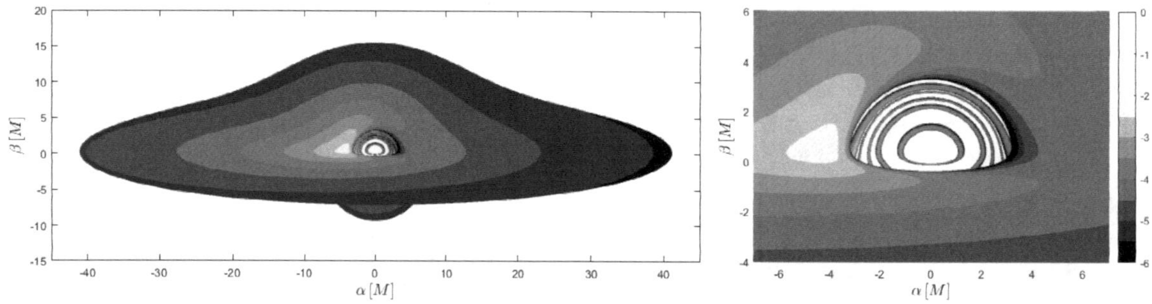

Figure 2. Image of an accretion disk starting from $r_{in} = M$ and ending at $r_{out} = 40M$, for scaling parameter $\gamma = 1$ and observer located at infinity with inclination angle $i = 80$ deg. The color-map indicates the observed flux normalized to its maximum value in a logarithmic scale. The axes are in units of ADM mass.

5. Conclusion

We investigate the formation of characteristic features in accretion disk images around certain wormhole spacetimes. We present the results of an analytical and numerical treatment of the problem of image formation and find qualitatively new features of such spacetimes, which are altogether absent for black holes. This provides a potential avenue for the observational confirmation of the existence of such objects.

Acknowledgements

We gratefully acknowledge support by the Bulgarian National Science Fund Grant KP-06-H38/2 from December 5, 2019.

References

[1] Kelly P et.al. 2015 Multiple images of a highly magnified supernova formed by an early-type cluster galaxy lens. *Science*, **347**(6226):1123–6
[2] Tsapras Y 2018 Microlensing searches for exoplanets. *Geosciences (Switzerland)*, **8**(10),
[3] Akiyama et al. 2019 First M87 Event Horizon Telescope Results. I. The Shadow of the Supermassive Black Hole. *The Astrophys. J.*, **875**(1):L1
[4] Müller T 2009 Analytic observation of a star orbiting a Schwarzschild black hole. *Gen. Rel. Gravit.*, **41**(3):541–58
[5] Gyulchev G, Nedkova P, Vetsov T, and Yazadjiev S 2021 Image of the thin accretion disk around compact objects in the Einstein–Gauss–Bonnet gravity. *Eur. J. Phys. C*, **81**(10):1–16
[6] Page D and Thorne K 1974 Disk-Accretion onto a Black Hole. Time-Averaged Structure of Accretion Disk. *Astrophys. J.*, **191**:499–506

Journal of Physics: Conference Series **2255** (2022) 012002 doi:10.1088/1742-6596/2255/1/012002

[7] Teo E 1998 Rotating traversable wormholes. *Phys.Rev D*, **58** 2, 024014
[8] Nedkova P, Tinchev V and Yazadjiev S 2013 Shadow of a rotating traversable wormhole. *Phys. Rev. D*, **88** 12, 124019
[9] Shaikh R 2018 Shadows of rotating wormholes. *Phys. Rev. D*, **98** 2, 024044
[10] Luminet J 1979 Image of a spherical black hole with thin accretion disk. *Astron. Astrophys.*, **75**:228–235
[11] Paul S, Shaikh R, Banerjee P, and Sarkar T 2019 Observational signatures of wormholes with thin accretion disks. *JCAP* **03** 055
[12] Schee J and Stuchík Z Appearance of Keplerian discs orbiting on both sides of a reflection-symmetric wormhole. arXiv2111.00750

Galactic cosmic rays above the Earth's atmosphere

Marusya Buchvarova

Space Research and Technology Institute-BAS
Acad. Georgi Bonchev Str., bl. 1, Sofia 1113, Bulgaria

Corresponding author: marusjab@yahoo.com

Abstract. Galactic cosmic rays (GCRs) originate from sources outside the solar system and reach the Earth's environment from all directions. More than 100 years after the first detection of cosmic rays, the origin of high-energy cosmic rays is still a mystery. Although our knowledge of the origin and propagation of cosmic rays is relatively limited, we can study the radiation conditions in the near-Earth space environment in more detail. The origin, composition, and energy spectra of cosmic rays are briefly discussed in this paper. A brief review of the transport of galactic cosmic rays in the heliosphere is given. The geomagnetic effects on galactic cosmic rays and the atmospheric interactions of primary particles are also discussed in the paper.

1. Introduction

The discovery of cosmic rays by Victor Hess in 1912 opened a new window for the matter in the Universe. The beginning of a whole new world of particles beyond the confines of the atom was set [1]. The first particle of antimatter, the positron, was discovered in 1932, the muon was discovered in 1936, followed by the pion, the kaon, and many others. In 1956, the Cowan–Reines neutrino experiment confirmed the existence of neutrinos; until then, the neutrino had been a purely hypothetical particle. Today, we know that galactic cosmic rays travel through interstellar space and the heliospheric magnetic field, leading to a strong modulation of their spectrum, and also pass through the Earth's magnetosphere before reaching us.

In the near-Earth space, the main sources of corpuscular radiation are galactic cosmic rays (GCRs), particles of solar origin as solar energetic particles (SEPs), and particles trapped in the Earth's magnetosphere (radiation belts). Before cosmic rays (CRs) interact in the atmosphere they are known as primary cosmic rays (PCRs). Although cosmic rays include galactic cosmic rays and other classes of energetic particles, including solar energetic particles and particles accelerated in interplanetary space, the term "cosmic rays" is often used to refer only to galactic cosmic rays.

Galactic cosmic rays are charged particles originating in sources outside the Solar system that travel through space at nearly the speed of light, and reach the Earth from all directions. They consist of ionized nuclei, mainly protons, some helium, and less than 1% heavier elements. Among the cosmic rays, other subatomic particles (besides protons) can also be found. The particles with intrinsic mass are known as "cosmic rays" while photons, which are quanta of electromagnetic radiation (and therefore have no intrinsic mass), are called by their common names such as gamma rays or X-rays, depending on their photon energy [2].

Cosmic rays can have a wide range of energies, from tens of MeV to 10^{20} eV. Particles with energies less than 30 MeV/nuc are often neglected in descriptions of space radiation environments because their range is so small that they would not pass through typical shielding levels [3]. The low-energy cosmic rays can only be detected at high latitudes and altitudes. Their spectra can be recorded onboard spacecraft, rockets, and balloons [4]. In the energy range above 1 GeV, CR spectra are well measured by ground-based neutron monitors and muon telescopes.

Cosmic ray particles are divided into two main classes depending on their energy and probable acceleration sites [5]: charged particles with energies less than 10^{18} eV are referred to as galactic cosmic rays; above that energy (~ 10^{18} eV), they are referred to as extragalactic cosmic rays or ultrahigh-energy cosmic rays. Although galactic cosmic rays are thought to be accelerated in the shock waves generated by powerful supernova explosions [5], the sources and sites of cosmic ray acceleration are still debatable. In addition, the origin of the highest-energy particles remains a mystery. By comparison, the most powerful man-made particle accelerator, the Large Hadron Collider near Geneva, reaches energies of 10^{13} eV, which is barely one ten-millionth of the energy of the registered cosmic rays with the highest energy [5].

The magnetic fields of our Galaxy have two different components: a regular and a turbulent one [6]. The strength and directions of the magnetic fields that belong to the turbulent component are random [6, 7, 8]. The charged particles are deflected by the magnetic fields in the galaxy, so there is virtually no connection between the directions from which the particles seem to come to us and the directions of their sources. The cosmic ray flux that arrives at the Earth is also random or, more precisely, isotropic [6]. For this reason, it is difficult to identify specific astronomical objects as sources of cosmic rays. The particle flux is not only isotropic but also approximately constant in time, so the GCRs create a nearly uniform background of ionizing radiation that hits the Earth's atmosphere. Due to extremely high energies, high-energy cosmic radiation is the most penetrating type of cosmic radiation in the Earth's atmosphere.

2. Origin and composition of galactic cosmic rays

Cosmic rays come from all over the Universe. Particles with energies up to ~10^{18} eV must be accelerated by galactic sources. Above that energy, their sources must have extragalactic origin, although the exact demarcation energy remains somewhat vague and debatable, and there could be some overlap [5]. Galactic nuclei are believed to be accelerated at supernova remnant (SNR) shocks via first-order Fermi mechanism and to be then somehow released into the ISM, where they reside for some time before escaping the Galaxy [9]. This is the so-called SNR paradigm for the origin of galactic cosmic rays [10]. However, some types of expanding supernova remnants, for example, are estimated not to be able to accelerate protons above energies in the range of 10^{15} eV [11]. Possible sources of protons with those energies are massive star clusters (can reach PeV energies) [12] as well as the Galactic Center region [13]. The question remains whether the acceleration mechanism working in the SNRs can increase the energy of the particles all the way to the PeV region and beyond. What is the capability of SNRs to accelerate galactic cosmic rays to these extreme energies is an open question that requires more in-depth research. Cosmic rays with energies above ~10^{18} eV are thought to be of extragalactic origin. Their sources are not known. Candidates for sources range from the birth of compact objects to explosions related to gamma-ray bursts or to events in active galaxies [14, 15].

According to Ginzburg & Syrovatskii (1964) [16]: "Any theory of the origin of cosmic rays cannot expect serious success unless it rests on a detailed analysis of the observed composition of primary cosmic radiation". All-natural elements from the periodic table are present in the composition of cosmic rays. The main features of the cosmic ray composition at low energies (< 10^{14} eV) are well- known, but at higher energies, they remain unclear [17]. It is generally believed that the cosmic ray "knee" at 10^{15} eV may reflect a gradual transition towards a heavier composition [18]. In the direct-measurement energy region, about 98% of the particles are protons and nuclei; ≈ 2% are electrons [19]. Of the protons and nuclei, ≈ 87% are protons (H), ≈ 12% are helium nuclei (He) and a small fraction of ≈ 1% are heavier nuclei [19]. GCR nuclei of elements whose charge Z is higher than that of helium (with a charge of +2) are called HZE ions. De Angelis and Pimenta note in [20] that the composition is expected to vary with energy; however, only the energies below the knee are responsible for the known proportions. Cosmic rays in the Galaxy consist of two components: a primary component, representing the composition of the source, and a secondary component (Li, Be, B, sub-iron elements) produced by the interactions of the primaries with interstellar gas. Note that "primary" and "secondary" are used in a different but analogous sense when discussing cosmic ray interactions in the atmosphere [11].

3. Cosmic ray energy spectrum

The primary cosmic-ray particles extend over at least 13 decades in energy, with a corresponding decline in intensity of over 32 decades [21]. The all-particle CR spectrum obtained by compiling data from various instruments is shown in figure 1.

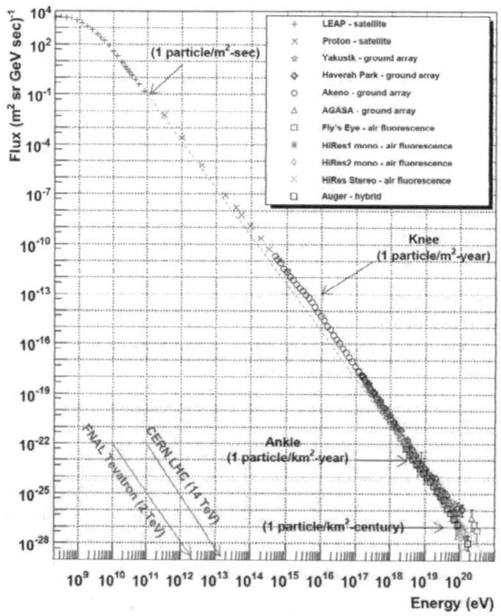

Figure 1. The overall differential energy spectra of cosmic rays from various experiments. Important features in the CR spectrum are indicated: the "knee" at 10^{15} eV and the "ankle" at 10^{18} eV. The frequencies of arrival of particles of different energies are pointed out, as well as the energies achievable in various accelerator experiments (Image from [22]).

The differential energy spectrum $D(E)$ is defined as the number of particles dN observed by a detector in the energy band from E to $E + dE$ passing through a unit surface area dA and arriving from a direction Ω within a solid angle $d\Omega$ during a time interval dt [6, 23]:

$$D(E) = \frac{dN}{dA.dt.d\Omega.dE} \tag{3.1}$$

Here, $D(E)$ is expressed in units of $m^{-2}\ s^{-1} sr^{-1}$ (MeV /nuc)$^{-1}$. As can be seen from figure 1, the GCR flux is a strongly decreasing function of energy. At first sight, the flux above a few GeV, where solar effects are negligible, can be approximated by a power law with a negative spectral index

$$D(E) \sim E^{-\gamma}, \tag{3.2}$$

γ is the spectral index of the cosmic ray flux. This part of the CR flux represents the spectrum beyond the heliosphere and is called the Local Interstellar Spectrum (LIS) [23]. The spectrum is almost smooth, with some slight deviations from a constant power law.

Below $E \sim 10^{15}$ eV, the power spectrum is a smooth curve with a spectral index of about 2,7. The energy spectra as a function of kinetic energy per nucleon of the primary ion components are very similar [24]. Below the knee, the statistics are still abundant enough and satellite and balloon experiments can provide information about the elemental composition of cosmic rays. In the conventional model for CR production, it is believed that the cosmic rays with the discussed energies are accelerated by astrophysical sources in our Galaxy, such as supernova remnants.

Two interesting features in the cosmic ray spectrum are observed: the “knee” and the “ankle”. Around the energy 10^{15} eV the spectrum becomes steeper, defining the ”knee”, and becomes flatter again above 10^{18} eV. The knee of the spectrum is visible as a change in spectral index from 2,7 to 3. The origin of this feature is still under study. It is generally believed that at energies above $\sim 10^{17\text{-}19}$ eV, the extragalactic CR particles dominate over the galactic ones. This seems to justify the hardening of the CR spectrum that defines another change in the slope, the so-called “ankle” [25, 26]. Here, the spectrum becomes less steep, and the spectral index changes back to $\gamma \approx 2{,}7$. It is believed that the threshold for the dominance of extragalactic sources is indeed close to the ankle [20]. In the energy interval between the knee and the ankle, the flux decreases significantly: there are only a few particles per square meter per year. Finally, at the highest energies, there is a drastic suppression of cosmic ray flux.

Cosmic rays with energies above 10^{18} eV are known as ultrahigh-energy cosmic rays (UHECRs). The flux of the primaries at these energies is extremely low, of the order of 1 particle per km^2 per century [27]. At the highest energies, cosmic rays are discovered by detectors with very large acceptance located on the surface of the Earth. The arrival rate of the most energetic particles is very low indeed, but particles with energies up to about 10^{20} eV have been detected [22]. At these energies, a cut-off in the spectrum is expected due to the interaction with the cosmic microwave background (known as Greisen - Zatsepin – Kuzmin (GZK) effect). The limit, called the GZK-limit, was computed in 1966 by Kenneth Greisen, Vadim Kuzmin, and Georgiy Zatsepin. They predicted that cosmic ray protons with energies above the threshold energy of 5.10^{19} eV would interact with cosmic microwave background photons to produce pions [21]. This would continue until their energy falls below the pion production threshold [21]. However, the cosmological sources and acceleration mechanisms for UHECR particles are still unclear.

Galactic cosmic rays after accelerated at their sources, suffer a combination of processes including diffusion, reacceleration, convection, nuclear interactions, radioactive decay, and energy losses. The most widely used models describing the propagation of cosmic rays in the interstellar medium are the diffusion, Leaky Box, and weighted slab models. These models aim to explain the cosmic ray propagation in the interstellar medium. Upon arrival in the solar system, the behavior of galactic cosmic rays, especially at lower energies, is affected by the solar wind and its embedded heliospheric magnetic field (HMF). For particles with higher energies, the heliosphere becomes transparent.

The charged particles with energies below 30 GeV will experience solar modulation (see Section 4). Figure 1 shows that the modulated spectrum peaks at energies of several hundred MeV/nuc [23]. The flux at these energies is high. For example, the rate of arriving particles at energies of about 10^9 eV is $\sim 10^4$ per square meter per second [21, 28]. Due to the interaction of galactic cosmic rays with the solar wind, our knowledge of the interstellar cosmic ray spectra at energies below a few GeV is obscured. Furthermore, because cosmic rays lose at least several hundred MeV/nuc when they traverse the inner heliosphere [29], we have essentially no information on the interstellar spectra of cosmic rays below ~200 to ~300 MeV/nuc [30]. Because of solar modulation and the fact that the nature of heliospheric diffusion coefficients is not yet fully established, all derived cosmic ray LIS at kinetic energies $T \lesssim 1$ GeV remain contentious [31].

4. Galactic cosmic rays in the heliosphere

Galactic cosmic rays (CRs) within the heliosphere interact with the solar wind and the heliospheric magnetic field embedded in it. This leads to significant global and temporal variations in their intensity and energy as a function of position inside the heliosphere [31]. The process is described as solar modulation [31]. The GCR modulation effect is strongest at the lowest observed energies and decreases

Journal of Physics: Conference Series **2255** (2022) 012003 doi:10.1088/1742-6596/2255/1/012003

with increasing energy, until it vanishes (within the uncertainties) at around 30 GeV [32]. For higher energy particles, the heliosphere is transparent. The GCR modulated flux is anti-correlated with solar activity, i.e. low cosmic ray intensities are observed during solar maximum periods and vice versa during periods of solar minimum.

The dominant and most important periodicity in the variation of CR intensity associated with solar activity is the 11-year cycle. This quasi-periodicity is convincingly reflected in the sunspot record since the early 1600s and also in the galactic CR intensity observed since the 1950s by a global network of ground-based neutron monitors [31]. Note that other solar activity indicators also vary with the 11-year periodicity, such as solar wind parameters, heliospheric magnetic field, flares and coronal mass ejections, and others. The 11-year periodicity in solar activity is referred to as the solar cycle. However, the 11-year sunspot cycle is half of a longer, 22-year cycle of solar activity. The Sun reverses magnetic polarity (the north magnetic pole becomes the south and vice versa) around the maximum of each 11-year solar cycle. Thus, the complete magnetic cycle is, on average, about 22 years - twice the duration of the sunspot cycle [33]. Besides these cycles, some scientists believe that there are other, longer-period variations in the sunspot numbers and solar cycles [33]. However, most scientists think we need more data, spanning longer periods of time, to definitively resolve this issue [33].

During the 11-year solar cycle, significant changes occur in the structure of the heliospheric magnetic field. Our Sun emits a continuous radial stream of supersonic plasma, the so-called solar wind with an embedded solar magnetic field that develops into the HMF [31, 32]. Because the HMF remains rooted at the photosphere of the rotating Sun, it forms an Archimedean or so-called Parker spiral [32, 34, 35]. Over the years, many modifications to this field have been proposed, but the realization of Fisk (1996) [36] that the differential rotation of the Sun and the rigid rotation of polar coronal holes have a significant effect on the structure of the HMF led to second-generation global HMF models that are too complex [31]. The main attribute of the HMF is that it follows a 22-year cycle with a reversal approximately every ~ 11 years at the time of extreme solar activity [31, 37]. When the ~ 11-year period is directed outward in the northern hemisphere and inward in the southern hemisphere, it is referred to as the $A > 0$ solar magnetic cycle, and vice versa for the $A < 0$ cycle (see figure 2 above).

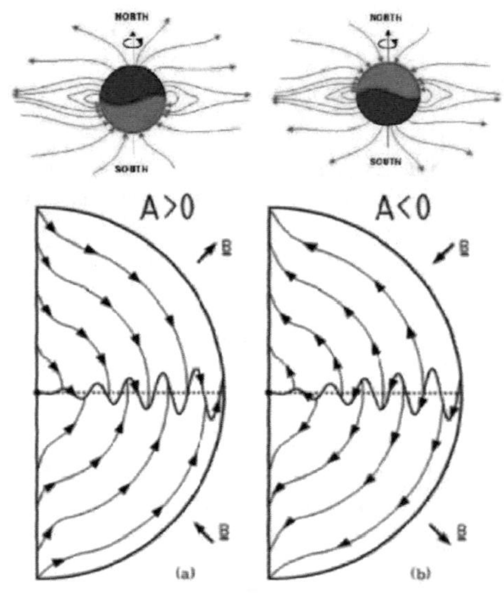

Figure 2. The idealistic picture of drift motion of positively charged particles at different polarities of the heliospheric magnetic field. During positive (left panel) /negative (right panel) polarity periods, the positively charged particles drift outward/inward along the heliospheric current sheet (Image from [31, 38]).

The largest structure of the HMF is the heliospheric current sheet (HCS). The HCS divides the heliosphere into regions with opposite polarities. The HCS demonstrates a wavy structure, parameterized by its tilt angle α (this is the angle between the dipole axis of the solar magnetic field and the rotation axis of the Sun). The tilt angle α determines the waviness of the HCS and correlates to solar activity. At high levels of solar activity, $\alpha = 75°$, but then becomes undetermined during times of extreme solar activity, while during solar minimum $\alpha = 3°–10°$ [31]. The waviness of the HCS has an important role in the CR modulation with the drift-dominated scenario, as discussed below.

The transport of cosmic rays in the heliosphere is formed by the combination of large-scale regular features and smaller-scale irregular fields [39]. The charged particles are scattered by irregularities in the heliospheric magnetic field, which is carried out with the solar wind. We describe the effects of this scattering as a diffusion process [40]. The magnetic field frozen into the solar wind tends to convect particles radially outward at the solar-wind speed. This sets up an outward gradient of cosmic rays that causes particles to diffuse inward [40, 41, 42]. In addition to the outward convection, the particles are subject to adiabatic cooling in the diverging solar wind [42]. The presence of a large-scale magnetic field affects the transport of particles. One effect of its presence is the existence of drifts [43]. Since there is a well-known difference in the cosmic ray modulation in successive polarity periods (see figure 3), it is important to consider particle drifts: gradient and curvature drifts as well as drift along the HCS [44]. Figure 2 shows idealistic global drift patterns for positively charged particles in A > 0 and A < 0 magnetic polarity cycles, together with a wavy HCS during solar minimum conditions [31].

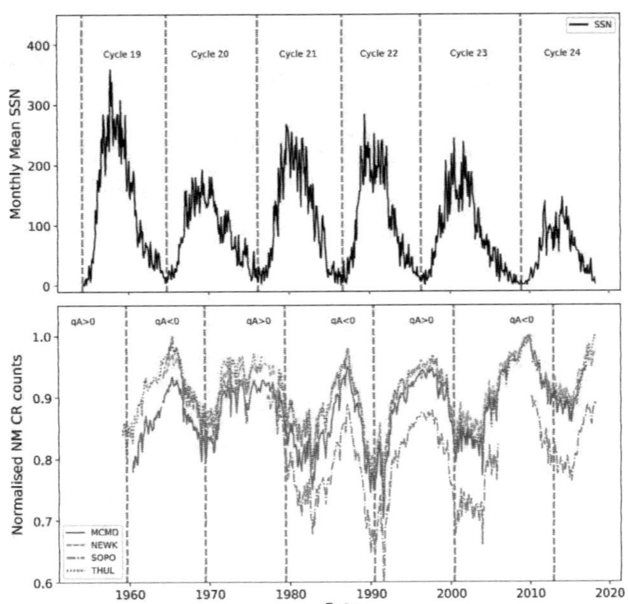

Figure 3. Monthly mean sunspot number (top), vertical lines indicate the beginning of each solar cycle. CR intensity recorded by NMs (bottom), vertical lines indicate the approximate epochs of solar magnetic-field polarity reversals (MCMD = McMurdo, NEWK = Newark, SOPO = South Pole, THUL = Thule) [45]. The time profiles of positively charged particles are more or less flat in the cycles with A>0 magnetic polarity like 1970s and 1990s, whereas in the cycles with A < 0 magnetic polarity the time profiles are peaked like 1980s and 2000s (Image from [45]).

During the A > 0 phase, when the field northward of the heliospheric current sheet (HCS) is directed outward, positively charged particles tend to drift towards the heliographic equator from the polar regions and drift outward along the HCS. When the polarity is reversed in the A < 0 phase, positively charged particles drift inward along the current sheet, and outward via the polar regions of the heliosphere. Negatively charged particles (electrons, antiprotons, all antimatter nuclei) show the opposite behavior. Therefore, the incoming GCR particles will be affected differently by the drift effects for the two magnetic configurations A>0 and A<0 [46]. As can be seen in figure 3, the 22-year solar magnetic cycle results in the alternating peaked and flat-topped shape of GCR intensity. The phenomenon is attributed to the "waviness" of the heliospheric current sheet (HCS) [47, 48]. Drifts may play an important role in cosmic ray transport during the period near solar minimum and 3–4 years after [39, 49]. However, it should be noted that perfect drift-dominated CR transport is not actually observed.

Although particle drifts are neglected in the simpler models of CR heliospheric modulation, more complex models explicitly take into account gradient, curvature and current sheet drifts in the HMF. Diffusion like propagation together with drifts, convection, and energy changes are the basic modulation mechanisms responsible for the 11- and 22-year variations in galactic cosmic ray intensity. Thus, the cumulative effects of the basic transport processes in the heliosphere are behind the solar modulation phenomenon of CRs [50]. The solar modulation is a space- and time-dependent phenomenon, i.e. it depends on where and when the CR flux is measured inside the heliosphere [50]. The spatial and temporal evolution of the GCR flux in the heliosphere can be described by the transport equation, introduced by Parker (1965) [51]. This equation is also known as the Parker transport equation.

5. Solutions of the cosmic ray transport equation

The Parker transport equation incorporates all modulation processes affecting the transport of GCRs inside the heliosphere [51]. In the heliocentric system, the Parker equation (also known as the cosmic ray transport equation) is expressed by [52]:

$$\frac{\partial U}{\partial t} = \frac{\partial}{\partial x_i}\left(K_{ij}^{S}\frac{\partial U}{\partial x_j}\right) - \frac{\partial}{\partial x_i}\left[\left(V_{\mathrm{sw},i} + v_{\mathrm{d},i}\right)U\right] + \frac{1}{3}\frac{\partial V_{\mathrm{sw},i}}{\partial x_i}\frac{\partial}{\partial T}\left(\alpha_{\mathrm{rel}}TU\right) \tag{5.1}$$

Here, $U \equiv U(r, T, t)$ is the differential number density of particles with kinetic energy T at time t at a heliocentric radius r. The solar wind speed along the axis x_i is given by $V_{\mathrm{sw},i}$, K_{ij}^{S} is the symmetric part of the diffusion tensor [51], v_{d} is the drift velocity, $\alpha_{\mathrm{rel}} = (T+2E_0)/(T+E_0)$ with E_0 – the particle rest energy. For non-relativistic particles $\alpha_{\mathrm{rel}}(T) = 2$ and $\alpha_{\mathrm{rel}}(T) = 1$ for extremely relativistic particles [43] so that $1 \le \alpha_{\mathrm{rel}} \le 2$ [53].

The cosmic ray transport equation for stationary modulation in a spherically symmetric solar wind with constant speed V, under the assumption that the diffusion is isotropic [54, 55] has a form:

$$\frac{\mathrm{V}}{r^2}\frac{\partial}{\partial r}\left(r^2 U_T\right) - \frac{1}{r^2}\frac{\partial}{\partial r}\left(r^2 k\frac{\partial U_T}{\partial r}\right) - \frac{2\mathrm{V}}{3r}\frac{\partial}{\partial T}\left(\alpha_{\mathrm{rel}}TU_T\right) = 0, \tag{5.2}$$

$U_T = U(r, T)$, $[U_\mathrm{T}]$ = particles/ (m^3 T) [43]. The terms in equation (5.2) describe, from left to right, convection, diffusion of particles and energy loss in the expanding solar wind. The number density U is related to the differential intensity D as [51]: $D = vU/4\pi$, v – the speed of the GCR particle.

The cosmic ray transport equation (TPE) is a partial differential equation of parabolic type [56]. In general, the equation cannot be solved analytically without simplifications. Due to its complexity, a numerical solution can be found when a realistic geometry of the heliosphere and a realistic functional form of the diffusion tensor are considered [57]. The first numerical solution of the TPE assuming a steady state and spherical symmetry was developed by Fisk (1971) [31, 55]. A brief history of the numerical models used to calculate cosmic ray intensities in the heliosphere can be found in [31, 58]. However, on the one hand, numerical modelling is a task that requires significant effort to accomplish [31]. On the other hand, the task concerning the GCR modulation has not been unambiguously solved theoretically because of insufficient knowledge of the properties of the modulation medium, the uncertainty of the boundary conditions (e.g., GCR energy spectra at the boundary of the modulation region), the time variations of these characteristics throughout the 11- and 22-year solar cycles [59, 60]. This motivates the use of simpler model solutions for practical purposes, such as the force-field and diffusion-convection approximations. In [56], Moraal (2013) summarizes the hierarchy of the model approximations in increasing order of complexity.

In 1969, Lennard Fisk and William Ian Axford published approximate analytical solutions for the cosmic ray transport equation in the heliosphere [61]. One of their solutions is used to develop useful semi-empirical models (such as the Matthiä model [62], the Kuznetsov model [63], the Nymmik model [59, 64]) capable of predicting galactic spectra during the solar cycle [65].

6. Geomagnetic effects on galactic cosmic rays

After their modulation in the heliosphere, the CRs reach the Earth almost isotopically. The last hurdle that CRs have to overcome before entering the Earth's atmosphere is the magnetosphere. Particles with high rigidities are able to pass through the magnetosphere, so there are no long-lived trapped particles with high rigidities. For low-rigidity particles, Kallenrode (2004) [66] distinguishes three cases: a) particles hitting the low-latitude magnetosphere from the outside perform half a gyro-orbit inside the magnetosphere before being reflected back into space; b) these particles can only penetrate into the magnetosphere at the polar cusps and interacting with the atmosphere produce polar cap absorption (PCA) events; c) trapped particles forming long-lived radiation belts.

The Earth's magnetosphere is a vast comet-like region around the planet, dominated by the Earth's magnetic field. Galactic cosmic rays are charged particles and the Earth's magnetic field affects their trajectories. Depending on the altitude above Earth' surface, the magnetosphere provides varying degrees of protection from solar and galactic particles by attenuating particle fluxes with atmospheric depth [67]. While the low-energy particles of the solar wind are deflected around the magnetosphere and are cut off, cosmic rays with energies in the tens of MeV and greater have direct access to the inner magnetosphere [68].

The rigidity R is the most appropriate physical quantity describing the effect of magnetic field on charged particles. The R is connected to the Larmor radius r_L in a field of magnitude $|\vec{B}|$ by

$$R = r_L\, |\vec{B}| c = cp/eZ \tag{6.1}$$

Because the energy (and pc) is expressed in eV, the charge in terms of the number Z of elementary charges, i.e. Ze, where $e = 1{,}602\times10^{-19}$ C, then R has units of Volts (V) [56]. CR particles with high rigidity (or energy) above a certain threshold will penetrate into the magnetosphere with nearly straight trajectories, while particles with intermediate rigidity will penetrate with a curved path, and the trajectories of low-rigidity particles will get strongly deflected such that they may even bounce back to their original direction of incidence [23]. This is the main reason why not all galactic particles that penetrate deep into the heliosphere reach the top of the Earth's atmosphere. Therefore, the intensity of galactic particles below certain energies, corresponding to the particle rigidity, is reduced [23]. It means that there is a screen effect for particles with a rigidity R below a threshold, called the cutoff rigidity Rc [69]. The charged particle cutoffs are a function of geomagnetic latitude, altitude, and the zenith and azimuthal directions and energy of the incident particle [67]. In the general case (except for geomagnetic cutoffs near the magnetic equator), the geomagnetic cutoff is ''diffuse'', i.e. at each position and in each direction there is an upper cutoff value R_U and a lower cutoff value R_L at the limits of an intermediate region, the so-called penumbral region [70]. Three regions can be distinguished. R_U is an upper cutoff value above which all particles have access to the point of interest. This is a region of high rigidities. R_L is a lower cutoff value below which particles cannot penetrate (forbidden regions). The region between the upper and lower cutoffs is an intermediate region, the so-called penumbra, where some cutoff values are allowed and others are not. Figure 4 illustrates trajectory-derived cutoff rigidities and the penumbral structure for the location at 40^0 latitude and 0^0 longitude.

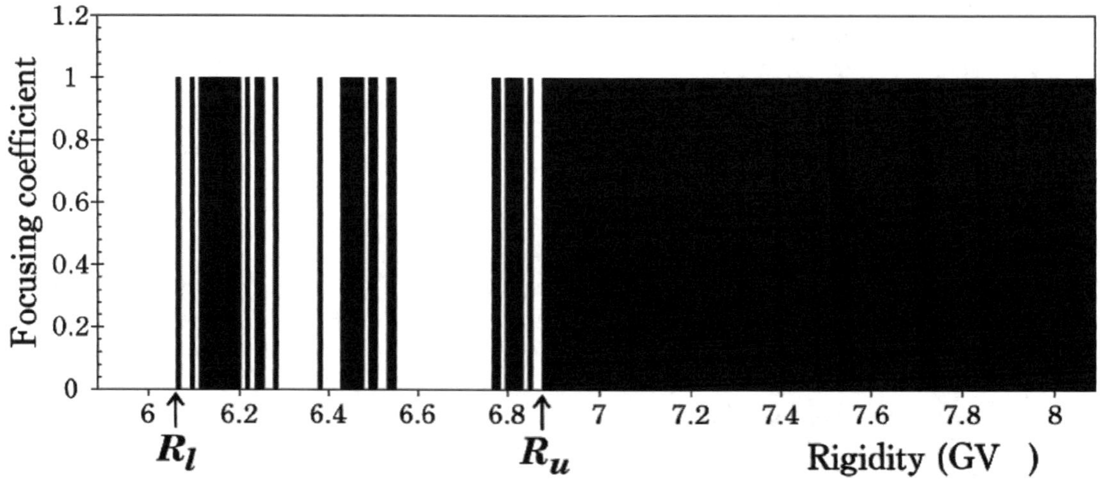

Figure 4. Illustration of the penumbra, i.e. the region of allowed and forbidden trajectories, calculated for the location: latitude = 40^0 and longitude = 0^0. The direction of the trajectory is zenith = 45^0 and azimuth = 0^0. The calculation is for a planetary radius of R_0 = 6448 km, i.e. for the top of the Earth's atmosphere. White regions indicate forbidden rigidities, and black regions indicate allowed rigidities. Focusing and dispersion is not considered (Image from [71], adapted for this work).

As can be seen from figure 4, the geomagnetic cutoff is not sharp. There is the penumbra, in which allowed and forbidden trajectories alternate [71]. More about the geomagnetic effects on galactic cosmic rays can be found in [71, 72] and many others.

For computational reasons, however, the penumbra is often approximated by an effective cutoff (R_{eff}) [71]. R_{eff} lies between R_L and R_U and takes into account the transparency of the penumbra. In calculations, the effective cutoff rigidity is considered as a hard cutoff: all rigidities above R_{eff} are allowed and all rigidities below R_{eff} are forbidden [71]. At low latitudes, near the equator, the penumbra is covered almost by forbidden regions, i.e. it is almost completely "dark", and at high latitudes, the penumbra is almost "transparent" [73]. The penumbra is most complex at middle latitudes: with many alternating bands of allowed and forbidden regions over a wide range of rigidity [73].

Since the calculation of a complete world grid of cutoff rigidities for a variety of directions takes a huge amount of time, the vertical cutoff approximation is the most widely used set of cutoff rigidities. The effective vertical cutoff rigidity, P_c, is a useful measure for comparing and interpreting particle measurements made at different locations on Earth, especially at different geomagnetic latitudes. The effective vertical cutoff rigidity P_c (often called vertical cutoff rigidity or simply cutoff rigidity) lies in the rigidity interval between the value below which (lower threshold) no particle arrives at the site of interest from the vertical direction and the highest rigidity value above which (upper threshold) all particles from that direction gain access. P_c depends strongly on the geomagnetic latitude, the altitude of the observation point, and the geomagnetic conditions at the time of observation [23]. The penumbral region is completely forbidden for vertical arrival between the equator and 20^0 geomagnetic latitude, but at higher latitudes alternate bands of allowed and forbidden rigidities are found and the penumbra becomes increasingly transparent [74].

The strength of the shielding effect of the geomagnetic field is greatest at the equator, where the field lines are parallel to the surface and the rigidity threshold is high. Here, all cosmic rays with energy less than ~15 GeV cannot penetrate the upper atmosphere [6] and the number of incident particles is correspondingly low. However, the rigidity threshold decreases with increasing latitude and becomes weaker and weaker [6]. After the 50th parallel (or below the - 50th parallel), the cosmic ray intensity is no longer dependent on the latitude and becomes constant [6]. The cutoff rigidities for ground-based cosmic ray stations and for vertical incidence are less than 1 GV near the magnetic poles and reach $P_c \approx 15$ GV at the geomagnetic equator [75].

The computations of the propagation of cosmic ray particles in the Earth's magnetosphere are performed using computer programs based on numerical integration of the equation of motion [75]. The program package MAGNETOCOSMICS, kindly provided by Desorgher (2004) [76] from the University of Bern, can be used to calculate the effective vertical cutoff rigidity in a grid covering the whole Earth.

Kress *et al.* [68] point out that there is very little variation in cutoff rigidity with the direction of arrival for high-latitude cutoffs and this justifies the use of vertical cutoff rigidities for modelling the *omnidirectional cosmic ray flux* in low Earth orbit. However, vertical cutoff rigidities are not appropriate for modelling cosmic ray flux falling on geosynchronous or other equatorial orbits from all directions [68]. At low geomagnetic latitudes, low-energy cosmic rays from the east are suppressed compared to those from the west. The effect, known as the east-west effect, demonstrates that positively charged particles arriving from eastern directions exhibit much higher cutoff rigidities than particles with western directions of incidence. The strength of the effect increases with increasing zenith angle [77]. The east-west asymmetry is greatest near the geomagnetic equator and decreases towards higher latitudes [78]. Note that the effect is strongest at the top of the atmosphere and weakens at sea level due to collisions with atmospheric nuclei.

In more detailed studies, the transmission of particles through the cosmic ray penumbra must be considered more precisely. In this case, Smart *et al.* (2006) [70] define a penumbral transparency, a quantity that describes the particle flux transmitted through the penumbral region. When the penumbral transparency is about 1, it shows that most of the particle flux in the penumbral region is allowed through the geomagnetic field. Penumbral transparency approaching zero indicates that a very small part of the particle flux in the penumbral region is allowed through the geomagnetic field [70]. Here, we would like to point out that there is a fundamental property of the CR flux in the allowed range of rigidities. Assuming an isotropic CR flux in interplanetary space, Liouville's theorem demands that for rigidities above the cutoff rigidity, the directional flux of particles in the magnetosphere is equivalent to that outside the influence of the geomagnetic field [68, 79, 80, 81].

The properties of cosmic ray transport from the Earth's magnetosphere boundary at 1 AU to an observation point inside the magnetosphere can also be described by the so-called transmission function (TF) (see Bobik *et al.* (2006) [82]). Bobik et al. (2006) [82] determine the TF for different geomagnetic regions of AMS-01 orbit. In the penumbral region, where the particle rigidity is almost equal to the geomagnetic cutoff rigidity, $0 < TF < 1$. TF = 0 in the forbidden region of rigidities, and TF = 1 for rigidities above the cutoff rigidity.

The concepts of cutoff rigidities and asymptotic directions were developed to describe geomagnetic effects on primary GCRs. An asymptotic direction is the direction of motion of a cosmic particle before it enters the magnetosphere. The asymptotic directions need not be considered when the cosmic ray intensity varies isotropically in interplanetary space. But, when the cosmic ray intensity varies anisotropically, the asymptotic directions must be considered.

The solar modulated cosmic ray fluxes which we measure on the top of the Earth's atmosphere, are also known as the top of atmosphere (TOA) fluxes [21]. At most ground-based CR experiments, where mostly vertically incident particles contribute to the counting rate, it is assumed that only CRs with rigidities greater than the effective vertical cutoff rigidity P_c can reach the top of the Earth's atmosphere from all directions of incidence. However, at high altitudes and in sites with high cutoff rigidity or in space, the contribution of non-vertical incident particles becomes significant, and the variation of cutoff rigidity with the direction must be taken into account [83, 84, 85]. The database: <http://lpsc.in2p3.fr/crdb> provides easy and efficient access to cosmic ray data from various balloon and space experiments. These data contribute significantly to the understanding of the global aspects of the modulation and allow the development of more accurate empirical and semi-empirical models describing the cosmic ray spectra during the solar cycle. One such semi-empirical model is discussed in [86] on this issue.

Journal of Physics: Conference Series **2255** (2022) 012003 doi:10.1088/1742-6596/2255/1/012003

Passing through the magnetic fields of the heliosphere and magnetosphere, GCRs strike the Earth's atmosphere. Primary CR particles with energies below several hundred MeV/nuc are simply stopped and absorbed in the atmosphere due to ionization losses [87]. However, if the energy of a primary particle is sufficiently high, it can interact with atmospheric nuclei and produce a cascade of secondary particles. In such collisions, many new particles are usually created and the colliding nuclei evaporate to a large extent [88]. During the development of the cascade (increase in the number of secondary particles), the flux of ionizing particles first increases down into the atmosphere, reaching its maximum [87] at altitude of 12–27 km (depending on season, solar activity level, and latitude), but then decreases with altitude due to the dominant absorption. Actually, on their way and in each interaction, the particles lose energy and eventually will not be able to generate new particles [88]. After some point, the shower maximum, more particles are stopped than generated, and therefore the number of shower particles decreases.

The altitude at which the flux of secondary particles is the largest is called the Pfotzer maximum. The height of the Pfotzer maximum varies with the geomagnetic latitude and solar cycle. Only a small fraction of the secondary particles usually comes down to the ground [88]. The number of secondary particles that actually reach the surface varies significantly depending on the energy and type of incident cosmic particles and the altitude (sea or mountain level). Atmospheric cascades initiated by higher-energy particles penetrate deeper into the atmosphere. However, for a given energy, heavier primary particles reach the shower maximum earlier, i.e. at shallower depths in the atmosphere than protons.

7. Conclusion

CR intensities and fluxes, as well as secondary particles that arrive on the ground, are of interest to scientists working in many different fields. It is believed that the beginning of the study of cosmic rays dates back to the pioneering balloon flights of Victor Hess, who measured the ionization of the atmosphere as a function of altitude. At the time, no one could have predicted that Victor Hess's flights would spawn new disciplines: astroparticle physics, high-energy astrophysics and elementary particle physics. Note that astroparticle physics is a relatively new field of research in which astrophysics, cosmology, cosmic rays, and particle physics work together to shed light on the nature and structure of matter in the universe [89]. It is often difficult to find a unique answer to the question "What is cosmic ray research?" This is because definitions can vary among the many professionals involved in cosmic ray research and because this field of research has encompassed many different areas of research throughout its long history [90]. In general, cosmic ray research involves the study of astrophysical systems, with scales ranging from the solar system, to our galaxy, to groupings of galaxies, focusing on the high energy processes that occur in these environments [90]. The galactic cosmic ray spectrum, at least for energies above 10^2 – 10^3 GeV/nuc, is fundamentally shaped by acceleration and diffusion; several other phenomena – e.g., convection, reacceleration, electromagnetic losses, and solar modulation – compete at lower energies [89]. The intensity of cosmic rays outside the Earth's magnetosphere depends on the local interstellar spectrum and its modulation in the heliosphere [91]. Note, however, that high-energy cosmic rays are not affected by the heliosphere.

Cosmic ray studies are not limited only to the traditional fields of high-energy physics and astronomy. Galactic cosmic rays are of interest to space weather researchers because they change radiation conditions in the near-Earth environment and can be considered an element of space weather. Furthermore, as dominant source of penetrating ionizing radiation, CRs affect physicochemical processes in the atmosphere. The energy input of CRs in the atmosphere is approximately equal to that of starlight and is about one-billionth of the solar irradiance [92]. However, cosmic ray ionization maintains the atmosphere as a very dilute electrically conducting plasma, allowing a continuous electrical current to pass from the ionosphere to Earth's surface [92]. The interaction of the CR particles with the atmosphere produces not only atmospheric ionization. The high-energy cosmic particles penetrating the dense layers of the atmosphere generate a rain of secondary cosmic rays, a small part of which continuously strikes the surface of the Earth. For example, about 20 primary particles arrive per square centimeter per second at the top of the atmosphere, while at sea level about one secondary cosmic ray particle arrives per square centimeter per minute [93]. In order to calculate the energy and number of secondary particles together with their direction of arrival and the time variations of these variables, the fluxes and intensities of the particles arriving at the top of the Earth's atmosphere must be precisely determined; the influence of the geomagnetic field and the CR interaction with the Earth's atmosphere must also be taken into account. Among natural radiation sources, cosmic radiation contributes to about 10-15% of the total annual effective dose received by the population at ground level [94,95].

References

[1] *CERN Accelerating Science* Cosmic Rays: Particles from Outer Space Available from: https://home.cern/science/physics/cosmic-rays-particles-outer-space
[2] *Cosmic Ray* Available from: https://en.wikipedia.org/wiki/Cosmic_ray
[3] Nelson G A 2016 *Radiat. Res.* **185** 349 Available from: https://bioone.org/journals/radiation-research/volume-185/issue-4/RR14311.1/Space-Radiation-and-Human-Exposures-A Primer/10.1667/RR14311.1.full
[4] Bazilevskaya G A 2000 *Space Sci. Rev*. **94**(1-2) 25
[5] Butt Y 2009 *Nature* **460** 701
[6] Ferrari F and Szuszkiewicz E 2009 *ASTROBIOLOGY* **9**(4) 413
[7] Rand R J and Kulkarni S R 1989 *Astrophys. J.* **343** 760
[8] Clay R W, Kurban Z and Wild N R 1998 *Cosmic Ray Related Undergraduate Experiments GAP Note 1998-061* Technical and Scientific Notes about the Pierre Auger Project (Pierre Auger Collaboration)
[9] Gabici S, Evoli C, Gaggero D, Lipari P, Mertsch P, Orlando E, Strong A and Vittino A 2019 *Int. J. Mod. Phys.* D, **28** (15) 1930022
[10] Siming Liu 2019 *Bulletin of the Chinese Academy of Sciences* **33**(3)
[11] Tanabashi M et al. 2018 (and 2019 update) *Phys. Rev.* D **98** 030001
[12] Mohrmann L *et al.* 2021 *PoS* **395** *(37th Int. Cosmic Ray Conf.* July 12th – 23rd, 2021 Berlin) 789
[13] Abramowski A *et al.* 2016 *Nature* **531** 476
[14] Kotera K and Olinto A V 2011 *Annu. Rev. Astron. Astrophys* **49** 119
[15] Walz D S 2016 *Constraining Models of the Extragalactic Cosmic-Ray Origin with the Pierre Auger Observatory* Ph.D. Thesis (Aachen: RWTH Aachen University)
[16] Ginzburg V I and Syrovatskii S I 1964 *The Origin of Cosmic Rays* (Oxford: Pergamon Press)
[17] Wu J 2012 *Measurements of Cosmic Ray Antiprotons with PAMELA and Studies of Propagation Models* Doctoral Thesis (Stockholm: Royal Institute of Technology)
[18] Schlaepfer H 2003 *Spatium* **11** (Association Pro ISSI) p 3 Available from: http://www.issibern.ch/PDF-Files/Spatium_11.pdf
[19] Ghia P L 2012 Cosmic ray detection *International School on AstroParticle Physics* (*Paris*) Available from: https://isapp2012paris.sciencesconf.org/conference/isapp2012paris/P_Ghia_CosmicRayDetection_I.pdf
[20] De Angelis A and Pimenta M 2018 *Introduction to Particle and Astroparticle Physics* (Cham: Springer)
[21] Putze A 2006 *Propagation of Cosmic Rays in the Earth's Atmosphere* Thesis (Grenoble: Joseph Fourier University at Grenoble)
[22] Longair M S 2011 *High Energy Astrophysics* (Cambridge: Cambridge University Press)
[23] Mrigakshi A I 2013 *Galactic Cosmic Ray Exposure of Humans in Space* Dissertation (Kiel: Kiel University)
[24] Esposito G 2002 *Study of Cosmic Ray Fluxes in Low Earth Orbit Observed with the AMS Experiment* PhD Thesis (Perugia: Università degli Studi di Perugia)
[25] Pacini A A 2017 *Rev. Bras. Ens. Fis.* **39**(1) e1306
[26] Bertaina M E 2014 *C R Phys* **15** 300
[27] Dayananda M A 2013 *Correlation Studies of Cosmic Ray Flux and Atmospheric and Space*

Weather Dissertation (Atlanta: Georgia State University)
[28] *Pierre Auger Observatory* Frequently Asked Questions Available from: https://www.auger.org/outreach/cosmic-rays/faq
[29] Goldstein M L, Fisk L A and Ramaty R 1970 *Phys. Rev. Lett.* **25** 832
[30] Mewaldt R A, Wiedenbeck M E, Scott L M, Binns W R, Cummings A C, Davis A .J, Israel M H, Leske R A, Stone E C and von Rosenvinge T T 2004 *AIP Conf. Proc.* **719** 127
[31] Potgieter M S 2013 *Living Rev. Sol. Phys.* **10** 3
[32] Gieseler J 2018 *Understanding Galactic Cosmic Ray Modulation: Observations and Theory* Dissertation zur Erlangung des Doktorgrades (Kiel: Faculty of Mathematics and Natural Sciences at Kiel University)
[33] *The Sunspot Cycle* UCAR Center for Science Education Available from: https://scied.ucar.edu/learning-zone/sun-space-weather/sunspot-cycle
[34] Owens M J and Forsyth R J 2013 *Living Rev. Sol. Phys.* **10** 5
[35] Parker E N 1958 *Astrophys. J.* **128** 664
[36] Fisk L A 1996 *J. Geophys. Res.* **101** 15547
[37] Petrovay K and Christensen U R 2010 *Space Sci. Rev.* **155** 371
[38] Laurenza M 2018 *14th Quadrennial Solar-Terrestrial Physics Symp.* 9-13 July Toronto
[39] Jokipii J R and Kóta J 2000 *Astrophys. Space Sci.* **274** (1-2) 77
[40] Fisk L A 1980 *Proc. Conf. Ancient Sun: Fossil Record in the Earth, Moon, and Meteorites* 16-19 October 1979 Boulder ed R. O. Pepin *et al.* (New York and Oxford: Pergamon Press) p 103
[41] Perko J S 1984 *Solar Modulation of Galactic Cosmic Rays: Techniques & Applications* Ph.D. Thesis (Durham: University of New Hampshire)
[42] Jokipii J R 1971 *Rev. Geophys.* **9**(1) 27
[43] Batalha L 2012 *Solar Modulation Effects on Cosmic Rays* Dissertation (Lisbon: Instituto Superior Técnico)
[44] Pei C, Bieber J W, Burger R A and Clem J 2012 *Astrophys. J.* **744**(2) 170
[45] Ross E and Chaplin W J 2019 *Sol. Phys.* **294**(1) 8
[46] Badruddin O P M Aslam 2012 *Sol. Phys.* **279** 269
[47] Kóta J and Jokipii J R 1983 *Astrophys. J.* **265** 573
[48] Zhao L-L, Qin G, Zhang M and Heber B 2014 *J. Geophys. Res.* **119**(3) 1493
[49] McKibben R B, Connell J J, Lopate C, Simpson J A and Zhang M 1995 *Space Sci. Rev.* **72**(1-2) 367
[50] Fiandrini E, Tomassetti N, Bertucci B, Donnini F, Graziani M, Khiali B and Reina Conde A 2021 *Phys. Rev.* D **104** 023012
[51] Parker E N 1965 *Planet. Space Sci.* **13** 9
[52] Bobik P *et al.* 2013 *AdAst* **2013** 793072
[53] Gleeson L J 1968 *Proc. Astron. Soc. Aust.* **1**(4) (Clayton Victoria: Monash University) p 130
[54] Cowsik R and Lee M A 1977 *Astrophys. J.* **216** 635
[55] Fisk L A 1971 *J. Geophys. Res.* **76**(1) 221
[56] Moraal H 2013 *Space Sci. Rev.* **176**(1-4) 299 Available from: https://doi.org/10.1007/s11214-011-9819-3
[57] Zhang M 2008 *ASP Conf. Ser.* **385** 63
[58] Manuel R 2013 *Time-dependent Modulation of Cosmic Rays in the Outer Heliosphere* Ph.D. Thesis (South Africa: North-West University)
[59] Nymmik R A, Panasyuk M I, Pervaya T I and Suslov A A 1994 *Adv. Space Res.* **14**(10) 759
[60] Nymmik R A and Suslov A A 1996 *Radiat. Meas.* **26**(3) 477
[61] Fisk L A and Axford W I 1969 *J. Geophys. Res.* **74**(21) 4973
[62] Matthiä D, Berger T, Mrigakshi A I and Reitz G 2013 *Adv. Space Res.* **51**(3) 329
[63] Kuznetsov N V, Popova H and Panasyuk M I 2017 *J. Geophys. Res.* **122**(2) 1463
[64] Nymmik R A, Panasyuk M I and Suslov A A 1996 *Adv. Space Res.* **17**(2) 19
[65] Buchvarova M 2021 *Bulg. J. Phys.* **48**(1) 42

[66] May-Britt Kallenrode 2004 *Space Physics: An Introduction to Plasmas and Particles in the Heliosphere and Magnetospheres* (Berlin Heidelberg: Springer-Verlag)
[67] Barth J L 1997 Modeling space radiation environments *Notes from the 1997 IEEE NSREC Short Course* (Piscataway NJ: IEEE Publishing Services)
[68] Kress B T, Hudson M K, Selesnick R S, Mertens C J and Engel M 2015 *J. Geophys. Res.* **120**(7) 5694
[69] Fiandrini E *et al.* 2016 *PoS* **236** (*34th Int. Cosmic Ray Conf.* 30 July- 6 August 2015 The Hague The Netherlands) 095
[70] Smart D F, Shea M A, Tylka A J and Boberg P R 2006 *Adv. Space Res.* **37**(6) 1206
[71] Hirtz J and Leya I 2021 Dispersion and focusing of cosmic rays in magnetospheres arXiv:2104.13019
[72] Vargas B and Valdés-Galicia J F 2011 *Proc. the 32th Int. Cosmic Ray Conf.* August 11-18 2011 Beijing China 1099
[73] Biswas S 2000 *Cosmic Perspectives in Space Physics* Astrophysics and Space Science Library (Dordrecht: Kluwer Academic Publishers)
[74] Quenby J J 1967 The time variations of the cosmic ray intensity *Kosmische Strahlung II / Cosmic Rays II* ed K Sitte *Handbuch der Physik/Encyclopedia of Physics* (Berlin· Heidelberg· Newyork: Springer-Verlag) p 310
[75] Bütikofer R. 2018 Cosmic ray particle transport in the Earth's magnetosphere *Solar Particle Radiation Storms Forecasting and Analysis* Astrophysics and Space Science Library ed O Malandraki and N Crosby (Cham: Springer) p 79
[76] Desorgher L 2004 User guide of the MAGNETOCOSMICS code *Technical Report* (Bern: University of Bern)
[77] Herbst K 2012 *Interaction of Cosmic Rays with the Earth's Magnetosphere and Atmosphere* Dissertation (Kiel: Kiel University)
[78] Bhattacharya P C 1942 *Proc. R. Soc. London* A **8**(2) 263
[79] Lemaitre G and Vallarta M S 1933 *Phys. Rev* **43**(2) 87
[80] Swann W F 1933 *Phys. Rev* **44**(3) 224
[81] Fermi E 1950 *Nuclear Physics* (Chicago IL: University of Chicago Press)
[82] Bobik P, Boella G, Boschini M J, Gervasi M, Grandi D, Kudela K, Pensotti S and Rancoita P G 2006 *J. Geophys. Res.* **111** A05205
[83] Clem J M, Bieber J W, Evenson P, Hall D, Humble J E and Duldig M 1997 *J. Geophys. Res.* **102** 26919
[84] Dorman L I, Danilova O A, Iucci N, Parisi M, Ptitsyna N G, Tyasto M I and Villoresi G 2008 *Adv. Space Res.* **42** 510
[85] Vainio R *et al.* 2009 *Space Sci. Rev.* **147** 187
[86] Buchvarova M and Draganov D 2022 Model of galactic cosmic ray spectrum above the Earth's atmosphere *J. Phys.: Conf. Ser.* this issue
[87] Mironova I A, Aplin K L, Arnold F, Bazilevskaya G A, Harrison R G, Krivolutsky A A, Nicoll, K A, Rozanov E V, Turunen E and Usoskin I 2015 *Space Sci. Rev.* **194** 1
[88] *Cosmic-ray Air Showers*
Available from: https://www.mpi-hd.mpg.de/hfm/CosmicRay/Showers.html
[89] Castellina A and Donato F 2013 Astrophysics of galactic charged cosmic rays *Planets, Stars and Stellar Systems* ed T D Oswalt and G Gilmore (Dordrecht: Springer) 725
[90] Wefel J P 1988 An overview of cosmic ray research: composition, acceleration and propagation *Genesis and Propagation of Cosmic Rays Nato ASI Series C: Mathematical and Physical Sciences* ed M M Shapiro and J P Wefel (Dordrecht: D. Reidel Publishing Company) 1
[91] Fichtner H, Heber B, Herbst K, Kopp A and Scherer K 2013 Solar activity, the heliosphere, cosmic rays and their impact on the Earth's atmosphere *Climate and Weather of the Sun-Earth System (CAWSES)* ed F-J Lübken (Dordrecht: Springer) 55
[92] Carslaw K S, Harrison R G and Kirkby J 2002 *Science* **298**(5599) 1732

Journal of Physics: Conference Series **2255** (2022) 012003 doi:10.1088/1742-6596/2255/1/012003

[93] Close F, Marten M and Sutton C 2004 *The Particle Odyssey: A Journey to the Heart of Matter* (Oxford: Oxford University Press)
[94] Radiation: Effects and Sources 2016 *United Nations Environment Programme* Available from: http://large.stanford.edu/courses/2017/ph241/kwan2/docs/unep-2016.pdf
[95] Cinelli G *et al.* 2019 *European Atlas of Natural Radiation* ed G Cinelli *et al.* (Luxembourg: Publications Office of the European Union)

Journal of Physics: Conference Series **2255** (2022) 012004 doi:10.1088/1742-6596/2255/1/012004

Model of galactic cosmic ray spectrum above the Earth's atmosphere

M Buchvarova and D Draganov

Space Research and Technology Institute - BAS
Acad. Georgi Bonchev Str., bl. 1, Sofia1113, Bulgaria

Corresponding author: marusjab@yahoo.com

Abstract. Galactic cosmic ray transport in the heliosphere is described by the well-known Parker transport equation. In 1969, Fisk and Axford [1] presented approximate analytical solutions to the cosmic ray transport equation. One of their solutions was later used to construct useful semi-empirical models describing the energy spectra of charged particles during the solar cycle. In this paper, a simplified model of the galactic cosmic ray spectrum is presented. The model can be used to describe the galactic spectra of protons and helium nuclei during the 11-year solar cycle in interplanetary space at about 1 AU.

1. Introduction

Cosmic rays (CR) are a form of ionizing radiation that originates in outer space. Cosmic rays include galactic cosmic rays and other classes of energetic particles, including solar energetic particles and particles accelerated in interplanetary space [2]. However, the term "cosmic rays" is often used to refer only to galactic cosmic rays, which originate in sources outside the solar system [2]. CR are an important source of ionization in the atmosphere; the CR ionization rates have a strong influence on fundamental atmospheric properties such as electrical conductivity, atmospheric chemistry, and the charging of cloud particles [3, 4]. The interaction of cosmic rays with the atmosphere leads not only to atmospheric ionization, but also to fluxes of secondary particles that affect the physicochemical properties of the terrestrial atmosphere. On Earth, we are protected from high-energy cosmic rays by the magnetosphere and the atmosphere. Unfortunately, there is no such protection in space. Dangerous space radiation comes from solar protons and galactic cosmic rays (GCR). Galactic radiation is one of the main sources of radiation exposure in interplanetary space, low Earth orbit and at aviation altitudes [5]. Increasing the thickness of spacecraft walls would be enough to protect astronauts from low-energy particles from the Sun, but high-energy galactic cosmic rays would interact with the shielding materials to produce even more radiation [6]. For this reason, it is important to predict what radiation astronauts will be exposed to in a spacecraft or on the International Space Station (ISS). Predicting GCR intensities at 1 AU requires the use of models that calculate the energy spectra of charged particles (protons and heavy charged particles) as a function of time [7]. Deriving accurate and applicable models of GCR fluxes is also important for studying the role of cosmic rays in atmospheric processes. In this paper, we present a simplified model of the galactic cosmic ray spectrum above the Earth's atmosphere (outside the magnetosphere).

2. Fisk and Axford's approximate solution of the cosmic ray transport equation

The transport and propagation of galactic CRs in the heliosphere is determined by four major processes: convection and adiabatic energy losses caused by the expanding solar wind, diffusion caused by irregularities in the global heliospheric magnetic field (HMF), and particle drifts caused by gradients and curvatures in the HMF and by the wavy heliospheric current sheet (HCS) [8]. These major processes were incorporated by Parker [9] into the cosmic ray transport equation.

For a radial wind with constant speed V and radial diffusion with isotropic diffusion coefficient, Parker's transport equation [9,10] has the form:

$$\frac{\partial U}{\partial t}+\frac{\mathrm{V}}{r^2}\frac{\partial}{\partial r}(r^2U)-\frac{2\mathrm{V}}{3r}\frac{\partial}{\partial T}(\alpha_{\mathrm{rel}}TU)-\frac{1}{r^2}\frac{\partial}{\partial r}\left(r^2k\frac{\partial U}{\partial r}\right)=0\,, \tag{1}$$

where $U\equiv U(r, T, t)$ is the differential number density (number of particles per unit volume with kinetic energy T) at time t, r is the heliocentric radius. The solar wind speed is given by V, the particle diffusion coefficient by k, and $\alpha_{\mathrm{rel}}=(T+2E_0)/(T+E_0)$ with E_0 - particle rest energy (α_{rel} —troublesome factor [11]). For non-relativistic particles $\alpha_{\mathrm{rel}}(T)=2$ and $\alpha_{\mathrm{rel}}(T)=1$ for extremely relativistic particles [12], so that $1 \le \alpha_{\mathrm{rel}} \le 2$ [13]. On average, the changes in the heliosphere over the solar cycle occur so slowly with respect to the transit time of galactic cosmic rays and the solar wind through it that the time-derivative in equation (1) can be neglected [14]. Then, the cosmic ray transport equation for stationary modulation in a spherically symmetric solar wind with constant speed V, assuming that the diffusion is isotropic [10, 15] has the form:

$$\frac{\mathrm{V}}{r^2}\frac{\partial}{\partial r}(r^2U_T)-\frac{1}{r^2}\frac{\partial}{\partial r}\left(r^2k\frac{\partial U_T}{\partial r}\right)-\frac{2\mathrm{V}}{3r}\frac{\partial}{\partial T}(\alpha_{\mathrm{rel}}TU_T)=0 \tag{2}$$

Here, $U_T= U(r, T)$, $[U_T]$ = particles/(m^3 T) [12]. The terms in equation (2) describe, from left to right, convection, diffusion of particles, and energy loss in the expanding solar wind.

An alternative theory of cosmic ray propagation, developed by Axford (1965) [16], considers the Boltzmann equation of cosmic ray gas in the interplanetary magnetic field [17]. Gleeson and Axford (1967) [18] subsequently independently re-derived the transport equation for cosmic rays starting with a Boltzmann type equation for radial geometry. The CR transport equation derived by Gleeson and Axford (1967) [18] has the form:

$$\frac{1}{r^2}\frac{\partial}{\partial r}(r^2S_T)=-\frac{1}{3}\mathrm{V}\frac{\partial^2}{\partial r\partial T}(\alpha_{\mathrm{rel}}TU_T)\,, \tag{3}$$

where $S_T=S(r, T)$ is the radial differential current density or streaming

$$S_T=\mathrm{V}U_T-k\frac{\partial U_T}{\partial r}-\frac{\mathrm{V}}{3}\frac{\partial}{\partial T}(\alpha_{\mathrm{rel}}TU_T) \tag{4}$$

Eliminating the streaming S_T between equation (3) and equation (4), we obtain the cosmic ray transport equation (2) for the differential number density U_T. In deriving equation (2) Parker (1965) [9] used the concept of adiabatic deceleration, while Gleeson and Axford (1967) [18] used the concept of radially moving magnetic "scatterers" in deriving equations (3) and (4) [19].

Assuming that V and α_{rel} are constants, $k=\mathrm{k}_0T^{\mathrm{a}}r^{\mathrm{b}}$, ($a$, b - parameters, $b > 1$), and that $U(r, T)\rightarrow AT^{-\gamma}$ as $r \rightarrow \infty$, Fisk and Axford (1969) [1] show that when $S_T\approx 0$ equation (4) can be solved directly for the number density U_T:

$$U_T=AT^{-\gamma}\left(1-\frac{\mathrm{a}\alpha_{\mathrm{rel}}}{3(1-\mathrm{b})}\frac{\mathrm{V}r}{\mathrm{k}_0T^{\mathrm{a}}r^{\mathrm{b}}}\right)^{-\frac{\left[1+\frac{\alpha_{\mathrm{rel}}(\gamma-1)}{3}\right]}{\frac{\mathrm{a}\alpha_{\mathrm{rel}}}{3}}} \tag{5}$$

or

$$U_T = AT^{-\gamma}\Psi(T), \quad \Psi(T) = \left(1 - \frac{\mathrm{a}\alpha_{\mathrm{rel}}}{3(1-\mathrm{b})}\frac{\mathrm{V}r}{\mathrm{k}_0 T^{\mathrm{a}} r^{\mathrm{b}}}\right)^{\frac{\left[1+\frac{\alpha_{\mathrm{rel}}(\gamma-1)}{3}\right]}{-\frac{\mathrm{a}\alpha_{\mathrm{rel}}}{3}}} \tag{5a}$$

The form of the function $\Psi(T)$ in the above equation is used by several research groups [5, 7,20, 21, 22] for the construction of semi-empirical models that give the energy spectra of galactic cosmic rays as a function of time [23]. A function with the same form is also embedded in the model proposed below.

3. Semi-empirical model of galactic cosmic ray spectrum during 11-year solar cycle

The galactic cosmic ray differential spectrum $D(T)$, taking into account the changes in solar activity, is given by the expression:

$$D(T) = D_0(T)\left(1 + \frac{\alpha}{T}\right)^{-\Delta}, \tag{6}$$

$D_0(T)$ is the unmodulated GCR spectrum (LIS) with respect to the particle kinetic energy T. α and Δ are model parameters that depend on the solar activity level.

The differential cosmic ray intensity $D(T)$ as a function of particle energy T is related to the differential number density U_T by [24]

$$D(T) = \frac{vU_T}{4\pi}, \tag{7}$$

v-speed of the GCR particle. Then,

$$\frac{D(T)}{D_0(T)} = \frac{U_T}{U_{0T}}, \quad U_{0T} = AT^{-\gamma} \tag{8}$$

From equation (8), using equation (7) and considering that $\beta = \frac{v}{c}$, (c-the speed of light), we obtain:

$$D_0(T) = \frac{vAT^{-\gamma}}{4\pi} = \frac{\beta c A T^{-\gamma}}{4\pi}, \tag{9}$$

Substituting $C = \mathrm{c}A/4\pi$ into the above equation, we obtain the following expression for LIS

$$D_0(T) = C\beta T^{-\gamma} \tag{10}$$

If we assume that in equation (5) a = 1 and r = 1 AU, and take into account equation (8), then equation (6) follows the same form as equation (5).

Taking into account expression (6), the model equation for the calculation of the GCR proton and helium fluxes at 1 AU can be represented as

$$D_{(\mathrm{p,He})}(T) = D_{0(\mathrm{p,He})}(T)\left(1 + \frac{\alpha_{(\mathrm{p,He})}}{T}\right)^{-\Delta_{(\mathrm{p,He})}} \tag{11}$$

Here $\alpha_{(\mathrm{p,He})}$, $\Delta_{(\mathrm{p,He})}$ are parameters that depend on the particle type (protons, helium nuclei) and on the solar activity level. The LIS spectra of the individual particle species i take the form

$$D_{0,i}(T) = C_i\beta_i T^{-\gamma_i}, \quad i = (\mathrm{p,He}) \tag{12}$$

C_i – parameter, and γ_i - power index; C_i and γ_i depend only on the type of charged particles (protons, helium nuclei). $\beta_i = \sqrt{T(T+2m_i)}/(T+m_i)$, m_i is the rest mass of the particle; m_p=0,938 GeV for protons; m_{He}=0,939 GeV/nuc for helium nuclei [25].

The proposed model equation (11) describes the proton and helium spectra in the energy range from ~100 MeV to 100 GeV during the solar cycle.

4. Conclusion

Variations in intensity of cosmic rays are the consequence of the deflection of galactic cosmic rays by the time-varying solar magnetic field, which is convected away from the Sun by the solar wind [26]. The GCR intensity at Earth varies with time in response to solar activity, indicated by the sunspot number, with a period of ≈11 years; the amplitude of the variation increases with decreasing particle rigidity [26]. The transport of galactic cosmic rays in the heliosphere is well described by the cosmic ray transport equation [9]. In this paper, an approximate solution of Fisk and Axford (1969) [1] to the cosmic ray transport equation is discussed. This solution is embedded in a simplified model of the galactic cosmic ray spectrum. The model can be used to describe the proton and helium spectra of cosmic rays during the 11-year solar cycle in interplanetary space at a distance of about 1 AU. The fitting of the proposed model equation to experimental data and theoretical calculations will be done in the next work.

References

[1] Fisk L A and Axford W I 1969 *J. Geophys. Res.* **74** (21) 4973
[2] Mewaldt R A 1996 *Cosmic Rays* Available from: http://www.srl.caltech.edu/personnel/rmewaldt/cos_encyc.html
[3] Aplin K L 2013 *Electrifying Atmospheres: Charging, Ionisation and Lightning in the Solar System and Beyond* (Dordrecht: Springer)
[4] Nordheim T A, Dartnell L R, Desorgher L, Coates A J and Jones G H 2015 *Icarus* **245** 80
[5] Matthiä D, Berger T, Mrigakshi A I and Reitz G 2013 *Adv. Space Res.* 51(3) 329
[6] Archer-Boyd A 2014 Radiation shielding to protect a mission to Mars *Horizon, the EU Research and Innovation magazine* Available from: https://ec.europa.eu/research-and-innovation/en/horizon-magazine/radiation-shielding-protect-mission-mars
[7] Kuznetsov N V, Popova H and Panasyuk M I 2017 *J. Geophys. Res.* **122**(2) 1463
[8] Potgieter M and Nndanganeni R 2017 *PoS* **301** (*35th Int. Cosmic Ray Conf.* 10-20 July 2017 Bexco Busan Korea) 043
[9] Parker E N 1965 *Planet. Space Sci.* **13** 9
[10] Fisk L A 1971 *J. Geophys. Res.* **76**(1) 221
[11] Moraal H and Potgieter M S 1982 *Astrophys. Space Sci.* **84**(2) 519
[12] Batalha L 2012 *Solar Modulation Effects on Cosmic Rays* Dissertation (Lisbon: Instituto Superior Técnico)
[13] Gleeson L J 1968 *Proc. Astron. Soc. Aust.* **1**(4) (Clayton Victoria: Monash University) p 130
[14] Moraal H 1976 *Planet. Space Sci.* **19** 845
[15] Cowsik R and Lee M A 1977 *Astrophys. J.* **216** 635
[16] Axford W I 1965 *Planet. Space Sci.* **13**(2) 115
[17] Bemalkhedkar M M 1974 *Studies in cosmic rays* PhD Thesis (Ahmedabad: Physical Research Laboratory)
[18] Gleeson L J and Axford W I 1967 *Astrophys. J.* **149** 115
[19] Fisk L A, Gleeson L J and Axford W I 1969 *Acta Phys. Hungarica* (*Proc. the 11th Int. Conf. on Cosmic Rays* 25 August - 4 September 1969 Budapest **2** ed. A. Somogyi) Suppl. to **29** 105
[20] Nymmik R A, Panasyuk M I, Pervaja T I and Suslov A A 1992 *Nuclear Tracks and Radiation Measurements* **20**(3) 427
[21] Nymmik R A, Panasyuk M I, Pervaya T I and Suslov A A 1994 *Adv. Space Res.* **14**(10) 759
[22] Nymmik R A, Panasyuk M I and Suslov A A 1996 *Adv. Space Res.* **17**(2) 19
[23] Buchvarova M 2021 *Bulg. J. Phys.* **48**(1) 42

[24] Bobik P *et al.* 2013 *AdAst* **2013** 793072
[25] ISO 15390 2004 Space environment (natural and artificial)-Galactic cosmic ray model *International Standard*
[26] Caballero-Lopez R A, Engelbrecht N E and Richardson J D 2019 *Astrophys. J.* **883**(1) 73

Journal of Physics: Conference Series **2255** (2022) 012005 doi:10.1088/1742-6596/2255/1/012005

Observational features of hot spots on neutron stars with a photon sphere

Galin Gyulchev

Faculty of Physics, Sofia University, 5 James Bourchier Boulevard, 1164 Sofia, Bulgaria

E-mail: gyulchev@phys.uni-sofia.bg

Abstract. We describe and illustrate light deflection near neutron stars possessing a photon sphere motivated by modern confirmations of general relativity and its significance for relativistic astrophysics. In this paper, we model the spacetime outside the star with the Schwarzschild metric and summarize the properties of the photon orbits. More especially we illustrate the relativistic appearance of the hot spots on the neutron star surfaces over the celestial screen of an asymptotic observer. We visualize the relativistic images of the neutron star surface for various ratios of neutron star radius to Schwarzschild radius. We establish a noticeable increscent of the visual size of ultra-compact neutron stars with a photon sphere, which could distinguish stars with extremely strong surface gravity with respect to less compact neutron stars. We found an appearance of additional relativistic images of the hot spots leading to a higher magnification of the neutron star luminosity emission regions.

1. Introduction

Neutron stars are one of the most ultra-compact bodies in the Universe, possessing a highly relativistic stellar surface accessible to observations. A light ray emitted on or near the surface of a neutron star may be deflected by more than 90° degree in the case of ultra-compact neutron stars with a radius comparable to that of the photon sphere. Several authors have recently discussed the physics of this effect, the consequences for light curves from X-ray pulsars and considering the methods for calculation of the light curves [2, 3]. The foundations of the models for studying the visual effects of compact neutron stars have also been developed. In this context, Nemiroff analyses the relativistic distortions of the images near a neutron star and black hole, [4]. Some of these studies are even dedicated to the illumination of neutron star accretion discs, [5]. To explain the phenomena of rapidly-rotating neutron stars from an observational perspective, Olive also has investigated the effects of the mass quadrupole moment on the ray-tracing algorithm [6]. Recently, a paper by the NICER science team [7] has been published that describes a valuable method to extract a neutron star radius to mass ratio through the NICER's X-ray observations of pulsars with a rotational period of fewer than ten milliseconds [8]. The proposed method in the NICER experiment for extracting this so valuable information about the neutron stars is based on the relativistic gravitational lensing effect by highly mass density pulsars.

The study of the relativistic gravitational lensing when light passes very close to a massive body, such as an ultra-compact neutron star, has received wide attention recently. Nollert et al. have made a fundamental analysis of the relativistic "looks" of a neutron star [9]. Krauss [10] also makes a detailed qualitative analysis of light trajectories and study of images of neutron stars with different compactness. Thus we will restrict our research to an optical appearance of the effects of the light-bending by e highly compact neutron star possessing a photon sphere to give the reader an intuitive feeling for its astrophysical consequences.

2. Light Rays in the Schwarzschild Spacetime

We consider the spacetime metric outside and on the star's surface to be the Schwarzschild metric with mass M and radial coordinate r. The surface of the star is at $r = \mathcal{R}$ with $\mathcal{R}/2M \geq 1.25$ so the star is smaller than its "photon sphere" at $r = 3M$. As above, we will assume $G = c = 1$ throughout. Considering strongly relativistic neutron star possesses a photon sphere, we study the optical appearance of the stellar surface via numerical ray-tracing of the null geodesics $x^\mu(\lambda)$. The photon's four-momentum $p^\mu(\lambda) = dx^\mu/d\lambda$ satisfies the geodesic equation and is null [12, 13]:

$$\frac{dp^\mu}{d\lambda} + \Gamma^\mu_{\nu\sigma} p^\nu p^\sigma = 0, \qquad g_{\mu\nu} p^\mu p^\nu = 0. \tag{1}$$

We assume a stationary observer located at $r = r_{obs}$ in the asymptotically flat region of the spacetime far enough away from the neutron star. In the local inertial frame for any point of the trajectory parameterized by λ, the energy measured by the observer moving with four-velocity u^μ is [13]

$$\mathcal{E} = -g_{\mu\nu} u^\mu p^\nu. \tag{2}$$

The luminosity emission of the star is assumed to originate at $r = \mathcal{R}$ with specific intensity $I_\nu(\mathcal{R})$. The specific intensity $I_\nu(r_{obs})$ measured by the local observer is obtained via the photon number density $\mathcal{N}$, which is Lorenz invariant quantity and is constant along the null geodesic in vacuum [10, 11]

$$\frac{d}{d\lambda}\mathcal{N}(x^\mu(\lambda), p^\mu) = 0, \qquad \mathcal{N} = \frac{c^2}{h^4}\frac{I_\nu}{\nu^3}. \tag{3}$$

We also assume that the region under the stellar surface, $r < \mathcal{R}$, is entirely opaque to light rays, the region outside the surface, $r > \mathcal{R}$, is completely transparent, so we trace the photon trajectories near the surface without considering the presence of a stellar atmosphere.

2.1. The Geodesic Equation

Let us consider an observer with four-velocity u^μ who detect simultaneously two different photons with velocities v^μ and w^μ. In the rest frame of the observer the metric is $h_{\alpha\beta} = g_{\alpha\beta} + u_\alpha u_\beta$ and the photon velocities appears as the tree-vectors $v'^\mu = h^{\mu\alpha} v_\alpha$ and $w'^\mu = h^{\mu\alpha} w_\alpha$ [17, 18]. The observer sees angle α between the directions of the both photons, therefore the inner product of v'^μ and w'^μ the 3-space leads to

$$\cos(\xi) = 1 + \frac{(v^\mu w_\mu)}{(u^\nu v_\nu)(u^\sigma w_\sigma)}. \tag{4}$$

In the usual Schwarzschild coordinates t, r, θ and ϕ the Schwarzschild metric reads [13]

$$g_{\mu\nu} = \begin{pmatrix} -(1-\frac{2M}{r}) & 0 & 0 & 0 \\ 0 & (1-\frac{2M}{r})^{-1} & 0 & 0 \\ 0 & 0 & r^2 & 0 \\ 0 & 0 & 0 & r^2\sin^2(\theta) \end{pmatrix} \tag{5}$$

For the given static and spherically symmetric spacetime, the photon motion is planar and is described by the impact parameter, which is conserved quantity defined geometrically by the perpendicular distance from the center of the neutron star to the tangent of the light ray at the

Journal of Physics: Conference Series **2255** (2022) 012005 doi:10.1088/1742-6596/2255/1/012005

observer [21]. The impact parameter is an invariant of motion, which determines the precise shape of the orbit and represents the ratio between the photon's angular momentum $\mathcal{L} = p_\phi$ and the energy at infinity, $\mathcal{E} = -p_t$. For such a photon the geodesic parameter λ can be scaled through its energy, so that [19]

$$p^t \equiv \frac{dt}{d\lambda} = (1 - 2M/r)^{-1}, \tag{6}$$

$$p^r \equiv \frac{dr}{d\lambda} = [1 - (b/r)^2(1 - 2M/r)]^{-1}, \tag{7}$$

$$p^\theta \equiv \frac{d\theta}{d\lambda} = 0, \tag{8}$$

$$p^\phi \equiv \frac{d\phi}{d\lambda} = b\,r^{-2}. \tag{9}$$

In a particular situation of a radial motion the impact parameter is $b = 0$. Considering that special case the four-velocity of the photon is [15]

$$w^\mu = [(1 - 2M/r)^{-1}, 1, 0, 0]. \tag{10}$$

Letting the observer to be stationary its four-velocity in the Schwarzschild geometry is

$$u^\mu = [(1 - 2M/r)^{-1/2}, 0, 0, 0]. \tag{11}$$

Taking into account the Eqs. (10) and (11), as well as Eq. (4) for the angle measured on the observer's screen between the both vectors v^μ and w^μ we obtain

$$\cos(\xi) = [1 - (b/r_{obs})^2(1 - 2M/r_{obs})]^{1/2}. \tag{12}$$

It follows that

$$\sin(\xi) = (b/r_{obs})(1 - 2M/r_{obs})^{1/2}, \tag{13}$$

which in the limit $r \to \infty$ considering the photons falling into the observer reduce to

$$\xi = b/r_{obs}. \tag{14}$$

According Eq. (14) the quantity ξ presents the angular size of the neutron star's images visible to the asymptotic observer.

2.2. *The Trajectory*

Let us suppose that the photon emission from the surface of the neutron star ($r = \mathcal{R}$) is described by Eqs. (6) and obeys [12]

$$\left(\frac{1}{r^2}\frac{dr}{d\phi}\right)^2 + V(r) = \frac{1}{b^2}, \tag{15}$$

where

$$V(r) = (1 - 2M/r)/r^2 \tag{16}$$

is the effective photon potential. The Eq. (15) imposes restriction for a photon at r whose impact parameter must be such that

$$b^{-2} - V(r) \geq 0. \tag{17}$$

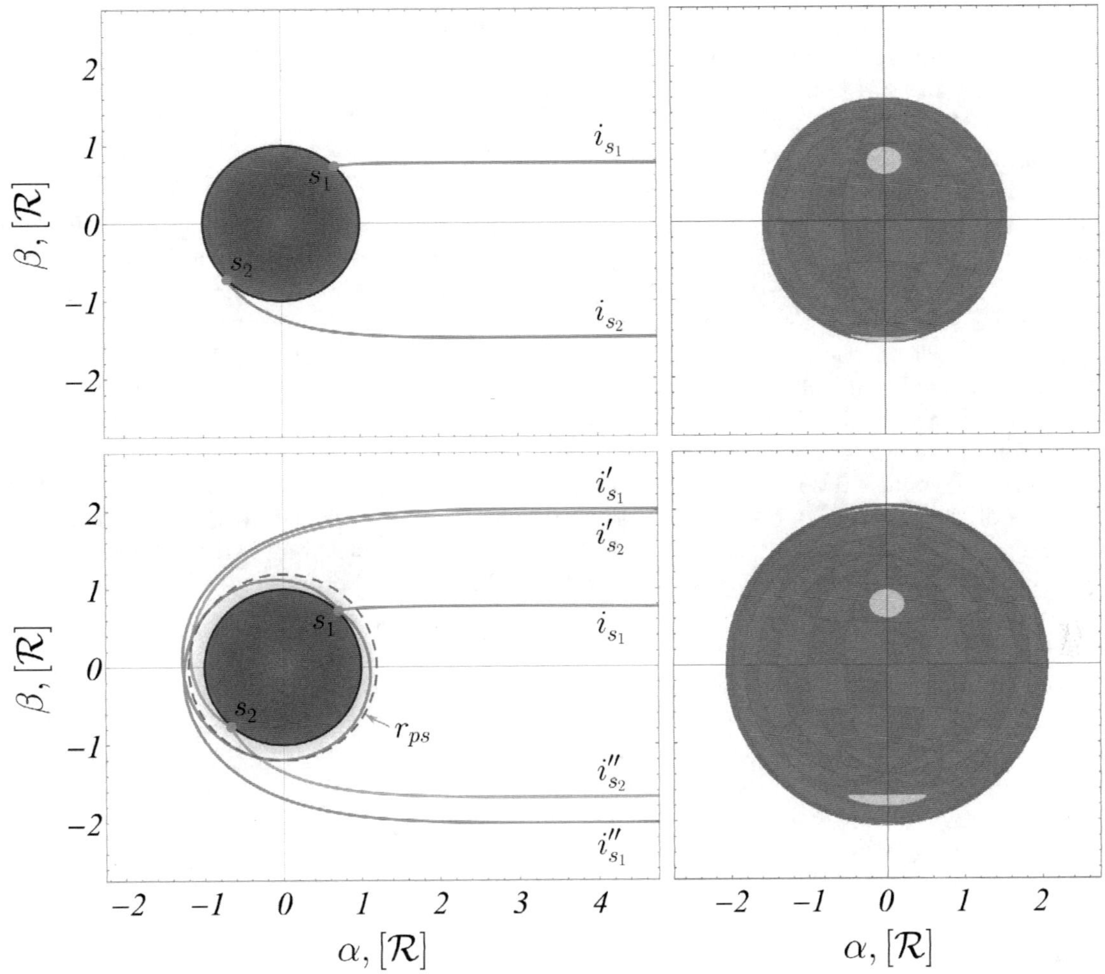

Figure 1. Photon's trajectories visualising the image creation of a hot spot on a neutron star with compactness $\mathcal{R}/2M = 1.7$ (*first row*) and formation of relativistic images of a hot spot on a neutron star possessing a photon sphere with compactness $\mathcal{R}/2M = 1.25$ (*second row*). The observer inclination angle is $\theta_{obs} = 90°$.

The effective potential has a maximum at the photon sphere, $r = 3M$, and decreases for $r > 3M$. Therefore, if the neutron star has radius $\mathcal{R} > 3M$, the photons must have an impact parameter $b \le b_{max}$ to reach the observer, where

$$b_{max} = \mathcal{R}(1 - 2M/\mathcal{R})^{-1/2}. \tag{18}$$

For a ultra-compact neutron star smaller than its photon sphere, a photon that starts tangentially to the stellar surface has a trajectory confined within the photon sphere and does not reach the observer. In order for the photons to leave the photon sphere its impact parameter must be smaller than the critical impact parameter [10]

$$b_c = r_{ps}(1 - 2M/r_{ps})^{-1/2} = 3\sqrt{3}M. \tag{19}$$

If the photon travels from $r = \mathcal{R}$ to $r \to \infty$ the trajectory is given by

$$\phi(r_{obs}) - \phi(\mathcal{R}) = \int_{\mathcal{R}}^{r_{obs}} \frac{b\,dr}{r^2\sqrt{1 - (b/r)^2(1 - 2M/r)}}. \tag{20}$$

Journal of Physics: Conference Series **2255** (2022) 012005 doi:10.1088/1742-6596/2255/1/012005

This integral is an elliptic integral of the first kind. With the Jacobi-sn function being the inverse of the Elliptic integral function the solution $r = r(\phi)$ can be obtained [20]. In our studies, we will solve this integral numerically.

3. Method

The deflection of light rays in the vicinity of a neutron star surface depends strongly on the ratio of the star's radius and its Schwarzschild radius. To achieve a sufficiently large deflection of light rays in the vicinity of the neutron star, we use a variety of compactnesses. Of particular interest is the value of $\mathcal{R}/r_S = 1.25$, at which the neutron star has sufficient mass per unit radius to possess a photon sphere. All obtained results are compared with those for the flat Minkowskian case characterized by compactness $\mathcal{R}/r_S = \infty$ as well as with the case $\mathcal{R}/r_S = 1.7$ when a photon sphere is absent.

We have created a specialized Mathematica code to solve numerically Hamilton's equations for the Schwarzschild metric. We are performing a back ray tracing of the light rays starting at the observer's screen and ending on the stellar surface with unit radius $\mathcal{R} = 1$. Figure 1 shows the paths of the photons as they travel form the neutron star's surface to the distant observer. All pictures have been computed with a resolution of 497×495 points, requiring around 500 sec of CPU time on a 6 core AMD Ryzen 5 PRO 4650U mobile processor.

4. Results

Figure 1 shows the path of the photons as they travel from the neutron star's hot spots to a distant observer. The light rays reach the observer with more significant impact parameters than they would in flat spacetime. Moreover, the photons emitted from the back half of the stellar surface can reach the observer because of the sufficiently strong gravity of the star. Moreover, some photons, which have impact parameters more negligible than the critical values b_c, typical for the photon sphere, can reach the observer, moving in the strong gravitational field of the star. This way, we can see the whole back half of the star surface in the case of sufficiently large compactness. The diagram mostly shows how the strong enough surface gravity of the ultra-compact neutron star is a reason for creating a different number of images i for each different location of the hot emission regions. Due to the presence of a photon sphere, the frontal hot spot has three images i_{s_1}, i'_{s_1} and i''_{s_1} instead of the back one, which possesses two images i'_{s_2} and i''_{s_2}. Therefore, we may see parts of the star that would otherwise be hidden from view.

Furthermore, the neutron star can create an infinite number of relativistic images of the whole stellar surface. Photons produce these images, making one or more complete loops around the neutron star's photon sphere before reemerging in the observer's direction. These relativistic images appear as an imprint of the stellar surface and could test the gravity near the star surface. Measuring the images' specific angular size, distribution, and deformation makes it is possible to determine the ratio between the neutron star's radius and its gravitational radius. In Figure 2 are shown all the different kinds of images of the neutron stars surface with various compactness for observer inclination angle. Studying the gravitational lensing by ultra-compact neutron star with a photon sphere, we are considering the particular case of a star with compactness $\mathcal{R}/2M = 1.25$ and compare those results with the case $\mathcal{R}/2M = 1.7$ as well as the Minkowskian case for $\mathcal{R}/2M = \infty$. Take a look at the significantly increasing visual stellar size and high elongation of the images of the hot spots for identical radii $\mathcal{R}$, but a growing mass M of the neutron star. One of the most valuable observational situations is the ability of the ultra-compact neutron star to create relativistic Einstein rings of its hot emission regions when the spot, centre of gravity of the star and observer are collinear. These images, shown in Figure 3, have circular symmetry and are incredibly bright, making them suitable for visually determining the apparent dimensions of the ultra-compact neutron star and its compactness.

5. Acknowledgements
The author gratefully acknowledge support by the Bulgarian National Science Fund Grants KP-06-H38/12 from 06.12.2019 and KP-06-RUSSIA/13 from 11.12.2020.

References
[1] F.W. Hehl, R.A. Puntigam, H. Ruder, Relativity and Scientific Computing: Computer Algebra, Numerics, Visualization, Springer, (1996).
[2] C. Fraclas, C. Kearney, M.W.K. Pechenick, Astrophys. J. **300**, 203 (1986).
[3] V. B. Bhatia, N. Chopra, B. Mujumdar, N. Panchakesan, Astrophys. J. **326**, 63 (1988).
[4] R. J. Nemiroff, Visual Distortions near a neutron star and black hole, *Am. J. Phys* **61**, 619-631 (1993).
[5] D. R. Wilkins, On the illumination of neutron star accretion discs, *MNRAS* **475**, 1, 748–756 (2018).
[6] G. A. Oliva, F. Frutos-Alfaro, Effects of the treatment of the mass quadrupole moment on ray-tracing applications for rapidly-rotating neutron stars, *MNRAS* **505**, 2, 2870-2885 (2021).
[7] Bogdanov et al, Constraining the Neutron Star Mass–Radius Relation and Dense Matter Equation of State with NICER. II. Emission from Hot Spots on a Rapidly Rotating Neutron Star, *ApJL* **887** L26 (2019).
[8] S. Bhattacharyya, Measurement of neutron star parameters: A review of methods for low-mass X-ray binaries, *Advances in Space Research*, **45**, 8, 949-978, (2010).
[9] H.P. Nollert, H. Ruder, H. Herold and U. Krauss, The relativistic "looks" of a neutron star, *Astronomy and Astrophysics* **208**, 1-2, 153-156 (1989).
[10] U. Krauss, Light Deflection Near Neutron Stars, *Relativistic Astrophysics*, 66-81, Vieweg+Teubner Verlag, Wiesbaden (1998).
[11] Bonning, Erin Wells. "Computational and astrophysical studies of black holes", PhD thesis, University of Texas at Austin, (2004)
[12] Dabrowski, M.P., Osarczuk, J. Light curves of relativistic charged neutron star. Astrophys Space Sci 229, 139–155 (1995)
[13] H. Riffert, H. Ruder, H.P. Nollert, F. W. Hehl, Relativistic Astrophysics, Springer Fachmedien Wiesbaden (1998).
[14] H. Sotani, Light curves from highly compact neutron stars with spot size effect, *Phys. Rev. D* **101**, 063013 (2020).
[15] K. R. Pechenick, C. Ftaclas and J. M. Cohen, Hot spots on neutron stars: The near-field gravitational lens, *ApJL*, **274**: 846-857 (1983).
[16] R. J. Nemiroff, P. A. Becker and K. S. Wood, Properties of ultracompact neutron stars, *ApJ* **406**: 590-595 (1993).
[17] G. F. R. Ellis, General Relativity and Cosmology, *Proceedings of the International School of Physics*, Course 47, ed R. K. Sachs (New York: Academic), 104-182 (1971).
[18] C. Ftaclas and J. M. Cohen, *Phys. Rev. D* **21**, 2013 (1980).
[19] C. W. Misner, K. S. Thorne and J. A. Wheeler, Gravitation (San Francisco: Freeman) (1973).
[20] T. Müller and J. Frauendiener, Interactive visualization of a thin disc around a Schwarzschild black hole, *Eur. J. Phys.* **33**:955 (2012).
[21] G.N. Gyulchev and S.S. Yazadjiev, Analytical Kerr-Sen dilaton-axion black hole lensing in the weak deflection limit, Phys. Rev. **D** 81, 023005 (2010).

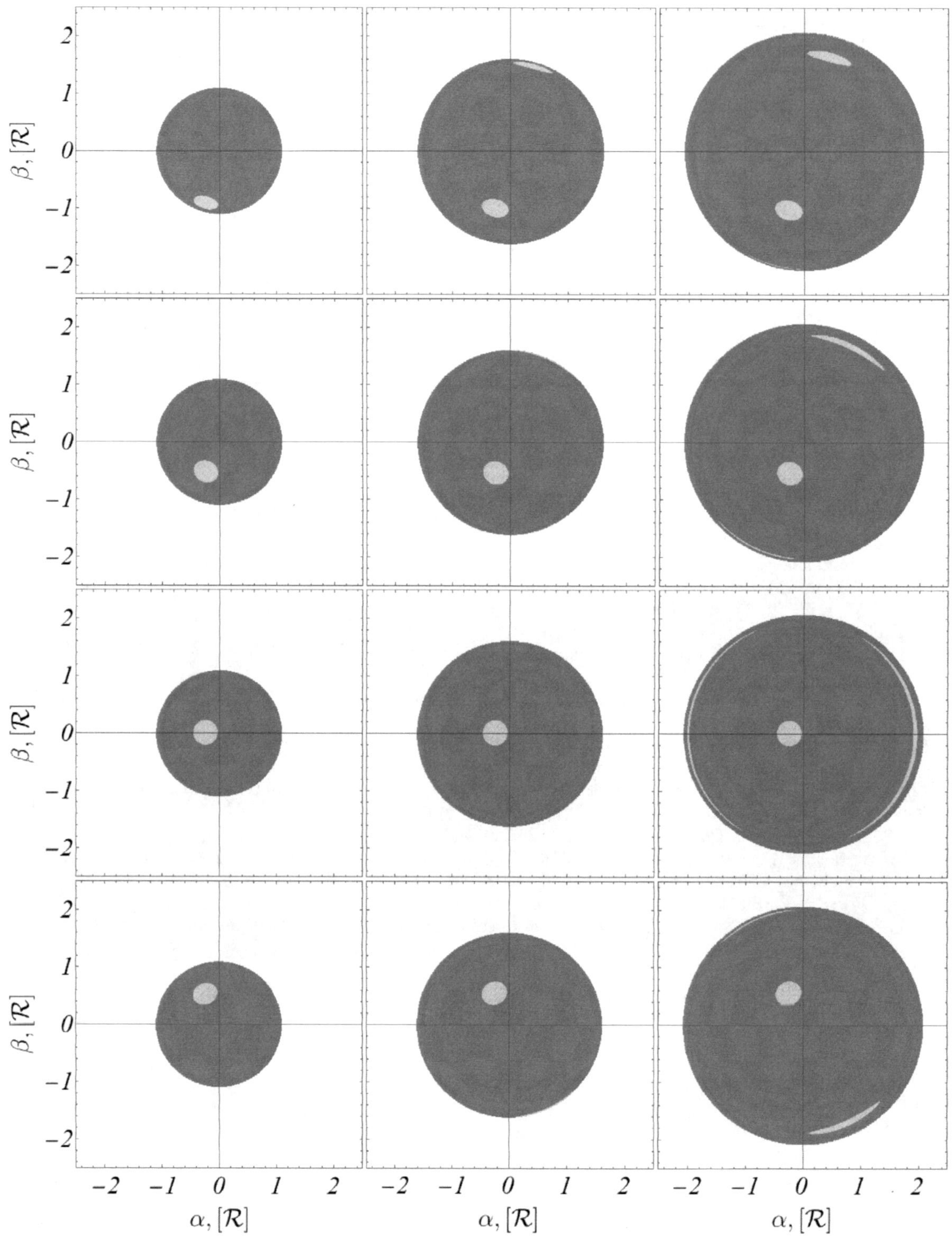

Figure 2. Image sequence of a neutron star with spherical coordinate grid for compactness $\mathcal{R}/2M = \infty$ (*left column*), $\mathcal{R}/2M = 1.7$ (*central column*) and $\mathcal{R}/2M = 1.25$ (*right column*) and observer inclination angle $\theta_{obs} = 0°$ (*first row*), $\theta_{obs} = 30°$ (*second row*), $\theta_{obs} = 60°$ (*third row*) and $\theta_{obs} = 90°$ (*fourth row*).

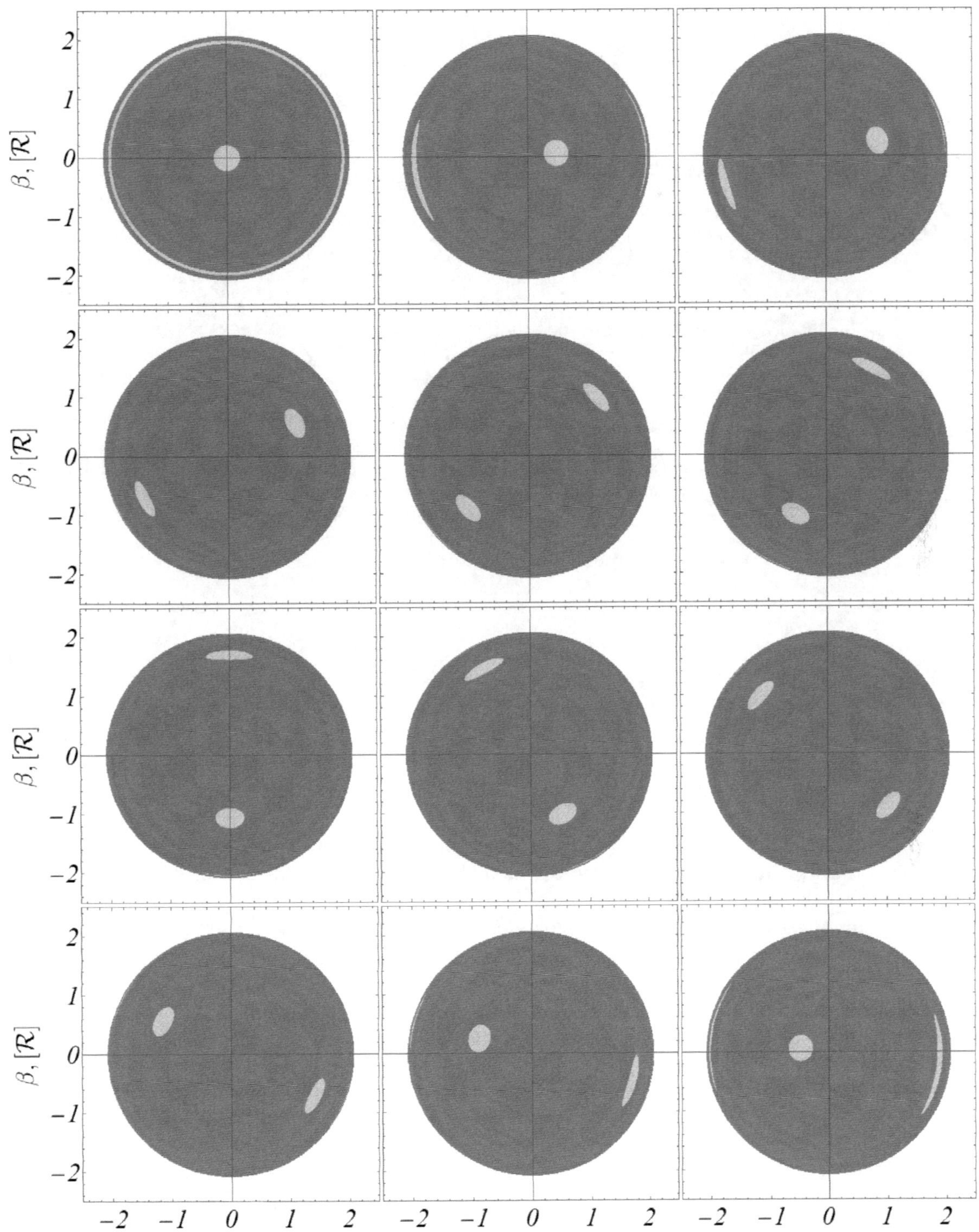

Figure 3. Image sequence of a neutron star with spherical coordinate grid for compactness $\mathcal{R}/2M = 1.25$ and various azimuthal angles $\phi = 0°, 30°, 60°, 90°, 120°, 150°, 180°, 210°, 240°, 270°, 300°$ and $330°$. The observer inclination angle is $\theta_{obs} = 60°$.

Coupled-cluster calculation of neutron matter equation of state

Stoyan Mishev

Institute for Advanced Physical Studies, 111 Tsarigradsko shose Blvd, 1784 Sofia, Bulgaria
New Bulgarian University, 21 Montevideo str. 1618 Sofia, Bulgaria

E-mail: stoyan.mishev@iaps.institute

Abstract. We present a relation between the density and pressure of pure neutron matter derived from numerical calculations based on a truncated coupled cluster model using a simple bare nucleon-nucleon interaction in the Minnesota potential form. The obtained relationship exhibits well-known behaviour only at low densities $\lesssim 0.08\ \mathrm{fm}^{-3}$ and diverges from the supported by observations data equations of state at higher densities. Our results are compared with those derived from effective field theory using realistic NN forces.

1. Introduction

From the conception of their existence in the early 1930s [1, 2, 3] the neutron stars have become very challenging objects for theoretical physicists from both the astrophysical and nuclear communities [4] as well as for the astronomers [5]. These are special places in the universe where the action of the gravity and the strong nuclear force have comparable magnitudes. The quest for understanding these dense compact stars is increasing rapidly in the last decade due to the previously unavailable data acquired from the instruments in the NICER, LIGO+Virgo and other facilities describing behaviours which provide us with additional important hints in resolving open questions such as the correct form of their equation of state. Among their hypothetical structural layers the inner cores of neutron stars hold most of the unknowns from the physics viewpoint mainly due to the onset of formation of a phase of matter in which the nucleons' constituents start to play more individually. From the other side, accelerators of heavy ions already provide direct characterization of this new phase of matter in the hot segment [6] and new experimental setups for exploring colder quark-gluon plasma are planned to start operation in the near future.

In this work we derive an equation of state of pure neutron matter using a microscopical approach for simulating the behaviour of systems of interacting particles which follows the coupled-cluster method (CCM) [7]. The ability of this method to produce accurate results based on pure NN potentials without introducing effective fields makes it suitable for establishing direct interconnections between the properties of the star reconstructed from observational data and the parameters of the inter-nucleon interaction. Moreover it allows to incorporate three- and four- nucleon forces which prove rather important for nuclear systems especially if compressed beyond the nuclear saturation density [8, 9, 10] . In the present paper however we limit our considerations with the two-body Minnesota nucleon-nucleon potential [11] which does not take into account the spin-orbit and tensor interaction. Furthermore we consider truncation of the

CCM wave function to double excitations and retained only pp and hh ladder terms in the Baker-Campbell-Hausdorff expansion of the similarity transformed two-body Hamiltonian as in Ref. [12]. In this respect the presented model can be considered only as a testbed for deriving a realistic equation of state from microscopic potentials by using the CCM techniques but it should be much improved. The equations of state obtained from various models must be able to reproduce available observational data such as their maximal masses [13], radii [14], mass-radius ratios, tidal deformability [15] and other in order to prove their robustness [16, 17]. In Ref. [18] a large ensemble of possible equations of state were exposed to constraints set by data on masses and tidal deformabilities to show that the surviving equations for maximally massive neutron stars and deformed as implied by LIGO exhibit patterns with two segments - one corresponding to the usual nuclear phase of matter and the other corresponding to a quark-gluon phase. Another perspective on the dense nucleonic regime is provided by the approach we use to understand the system as one of interacting clusters of particles held by many-body forces.

2. Nucleon correlations and coupled-cluster method

In this section we briefly show the formal steps to evaluate the ground state energy of a quantum system of N particles using a wave function in the exponential approximation as proposed by Coester and Kümmel [7] :

$$|\Psi\rangle = e^{T}|\Phi_0\rangle \tag{1}$$

in which the clustering operator acts on the Hatree-Fock (HF) state Φ_0 involving single ph ("singles"), double $2ph$ ("doubles") and higher excitations and reads

$$T = \sum_{ai} t_{ai} a_a^\dagger a_i + \frac{1}{4}\sum_{ai} t_{abij} a_a^\dagger a_b^\dagger a_j a_i + \dots = T_1 + T_2 + \dots \tag{2}$$

The ground state energy E then is the sum of the HF energy $E_{\rm HF}$ and the energy attributed to more complex configurations E_c :

$$E = E_{\rm HF} + E_c = \sum_i E_{ii} + \frac{1}{2}\sum_{ij} V_{ij,ij} + \langle\Phi_0|e^{-T} H_{\rm N} e^{T}|\Phi_0\rangle, \tag{3}$$

where E_i are the neutron energies and $H_{\rm N}$ is the normal-ordered Hamiltonian with respect to $|\Phi_0\rangle$

$$H_{\rm N} = \sum_{pq}\left(E_{pq} + \sum_i \langle pi|V|qi\rangle\right) : a_p^\dagger a_q : + \frac{1}{4}\sum_{pqrs} V_{pqrs} : a_p^\dagger a_q^\dagger a_s a_r : . \tag{4}$$

The structural coefficients t are the solution of the equations

$$\frac{1}{4}\sum_{ijab} t_{abij} V_{abij} = E_c \tag{5}$$

$$V_{abij} + \frac{1}{2}\sum_{kl} t_{abkl} V_{klij} + \frac{1}{2}\sum V_{abcd} t_{cdij} = 0, \text{ for each } a, b, i, j, \tag{6}$$

which are obtained by excluding the singles from the wave function and by considering only particle-particle and hole-hole ladder diagrams [12]. In the momentum conserving approach the basis states are suitably chosen as plain waves $\sim e^{i\vec{k}\vec{x}}$ in a cube L^3. In numerical

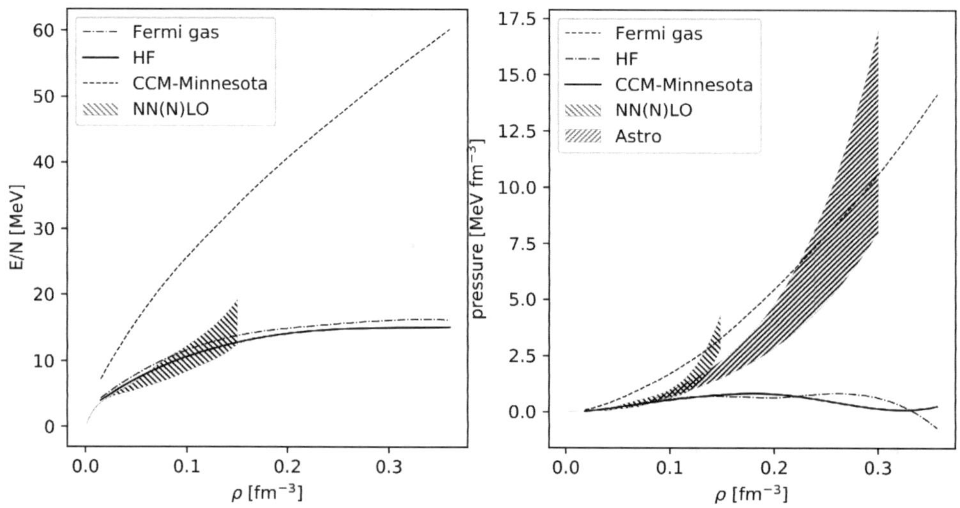

Figure 1. Left: Energy per neutron as function of the density calculated in the ideal Fermi gas approximation and also by adding HF correlations as well as more complex coupled-cluster correlations. The striped region shows the adapted results of the N^2LO and N^3LO calculations of Krüger *et al* [19] with inherent theoretical uncertainties. Right: Pressure as function of the density calculated using the same approaches as in the left panel. Additionally the observationally inferred relationship adopted from Ref. [16] is added.

calculations their number is limited by convergence arguments and is controlled by N_{max} so that $k_x, k_y, k_z = -\frac{2\pi}{L}N_{max}, \ldots, \frac{2\pi}{L}N_{max}$. The interaction strengths are thus evaluated in the momentum⊗spin basis as

$$V_{rspq} = \langle k_p s_p k_q s_q | V | k_r s_r k_s s_s \rangle. \tag{7}$$

3. Numerical results and discussion

We performed calculations at different densities using the Minnesota potential to obtain values for E_{HF}/N and E/N where N is the number of particles. The results are plotted in the left panel of Fig. 1a and compared with the energy of ideal Fermi gas of neutrons as well as with results of Krüger *et al* [19] involving effective field theory with chiral forces including next-to next-to-next-to-leading order (N^3LO) terms. In the right panel of this figure are plotted the corresponding equations of state of the neutron matter, i.e. the relationship between its pressure and matter density calculated as

$$P = \rho^2 \frac{\partial(E/N)}{\partial \rho}. \tag{8}$$

The right panel also shows the region of the supported by observational data equations of state[16] which contain other particle species beside the neutrons. One obvious conclusion is that the Minnesota potential cannot be considered as a realistic NN force for describing neutron stars in the applied here CCM approximation scheme as it lacks important ingredients namely the tensor and spin-orbit interaction. The chiral effective field theory calculations of neutron matter using three- and four- body forces on the other side produce results which are in tact with the data and our current efforts are focused in including more realistic potentials in the model.

4. Acknowledgments

The author thanks S. Yazadjiev and D. Doneva for the fruitful discussions during his stay at the University of Tuebingen where parts of this work were performed. The author acknowledges that this work was partially supported by the National Science Fund of Bulgaria, contract K-06–H-38/12 from 06.12.2019. We gratefully acknowledge the support of NVIDIA Corporation with the donation of the Titan V GPU used in this study.

References

[1] W. Baade, F. Zwicky, Proc. Natl. Acad. Sci. U.S.A. 20, 254–259 (1934)
[2] W. Baade, F. Zwicky, Proc. Natl. Acad. Sci. U.S.A. 20, 259–263 (1934).
[3] Landau L D Phys. Z. Sowjetunion 1 285 (1932)
[4] James M. Lattimer, Madappa Prakash, Physics Reports, Volume 442, Issues 1–6, April 2007, pp 109-165
[5] Bell Burnell, J. The past, present and future of pulsars. Nat Astron 1, 831–834 (2017)
[6] STAR Collaboration, Nuclear Physics A, vol. 757, Issues 1–2, 8 August 2005, Pages 102-183
[7] F. Coester, Nuclear Physics 7, 421–424 (1958); F. Coester and H. Kümmel, Nuclear Physics 17, 477–485 (1960)
[8] R. B. Wiringa, A. Arriaga, and V. R. Pandharipande, Phys. Rev. C 68, 054006 (2003)
[9] E. Epelbaum, H.-W. Hammer, and Ulf-G. Meißner Rev. Mod. Phys. 81, 1773 (2009)
[10] P. Navrátil, V. G. Gueorguiev, J. P. Vary, W. E. Ormand, and A. Nogga, Phys. Rev. Lett. 99, 042501 (2007)
[11] D. R. Thompson, M. Lemere, Y. C. Tang, Nuclear Physics A, Volume 286, Issue 1, 8 August 1977, Pages 53-66
[12] G. Hagen *et al*, Phys. Rev. C 89, 014319 (2014)
[13] James M. Lattimer, Annual Review of Nuclear and Particle Science, Vol. 62:485-515 (2012)
[14] J. Nättilä *et al* Astron. Astrophys. 608, A31 (2017)
[15] Jocelyn S. Read *et al*, Phys. Rev. D 88, 044042 (2013)
[16] Feryal Özel, Paulo Freire, Annual Review of Astronomy and Astrophysics 54, 401-440 (2016)
[17] James M. Lattimer, Madappa Prakash, Physics Reports, Volume 621, 21 March 2016, Pages 127-164
[18] Eemeli Annala *et al*, Phys. Rev. Lett. 120, 172703 (2018)
[19] T. Krüger *et al*, Phys. Rev. C 88, 025802 (2013)

Journal of Physics: Conference Series **2255** (2022) 012007 doi:10.1088/1742-6596/2255/1/012007

Some Warnings About Quantum Space Gravimetry Enhance Earth Observations Project

Plamen Fiziev

Sofia University Foundation for Theoretical and Computational Physics and Astrophysics, 5 James Bourchier Blvd., 1164 Sofia, Bulgaria
and
JINR, Dubna, 141980 Moscow Region, Russia

E-mail: `fiziev@phys.uni-sofia.bg`

Abstract. In this paper, we discuss in brief some basic issues of quantum space gravimetry, related to standard approach of geodesy which is based on the Newton model of gravity and Euclidean geometry. We emphasize the need to apply relativistic gravity in practical high-precision geodesy.

Here we do not intend to solve the existing hard experimental and theoretical problems, being essential for the topic: development of quantum gravity, physics of dark matter and dark energy, novel physical principles of extended general relativity, in particular, a nonlinear superposition principle in general relativity and its extensions, and so on.

Rather, we point out the fundamental unsolved problems, which are substantial for quantum space gravimetry and future practical high-precision geodesy. We outline the possible ways for their study and decision. Thus, to some extend, the present paper is a program for further developments, not a presentation of the final solutions.

Our goal is to warn corresponding scientific community about the ultimate necessity for going outside the frameworks of the formulated more than three century ago, and used up to now in geodesy, Newton gravity, together with Euclidian geometry.

At present, in the emerging high-precision geodesy one must replace them with modern models of gravity and corresponding non-Euclidean geometry.

Without using and further development of those issues, the interpretation of data obtained from high-precision measurements by satellites for geodetic use seems to be quite problematic, uncertain, and may be misleading for practitioners.

1. Introduction

Nowadays, the theory of gravity and its experimental verifications are under close scrutiny.

The main reasons are: Newton gravity, with its linear principle of superposition of gravitational fields of arbitrary sources, is not adequate in the case of modern high-precision-experiments and observations in celestial mechanics, astrophysics, cosmology, etc.

1. For many high-precision practical uses, the Newton model of gravity is replaced by General Relativity (GR), possibly with proper cosmological constant Λ.

2. On the one hand, GR describes very well a lot of phenomena at very different scales and we are still waiting for first indications of serious discrepancies between GR and observations and/or experiments.

3. On the other hand, it is obvious that GR cannot be the final theory of gravity, at least because after an almost a century of different intensive efforts and attempts, we are still unable

to make GR compatible with Quantum Mechanics (QM), as well as with some firmly established, but still not well-understood phenomena like Dark Matter (DM) and Dark Energy (DE).

4. There exists a striking discrepancy between high-precision data of measurements of the Newton constant G in various places on the Earth surface, see [1] and references therein, Fig.1, and also [2]. The cause may be in different systematic errors and different precisions of variety of devices, used for measuring of the Newton constant G.

The main unsolved question: "Is the Newton constant G really a constant or it is some kind of field?" must be solved experimentally by measurements on the Earth surface and in the space with the same mobile equipment, the same precision and the same systematic errors.

Note that one has to make difference between measurements of the Newton constant G, and the measurements of the gravitational acceleration g at given place, since for them one needs quite different experimental equipment [4, 5].

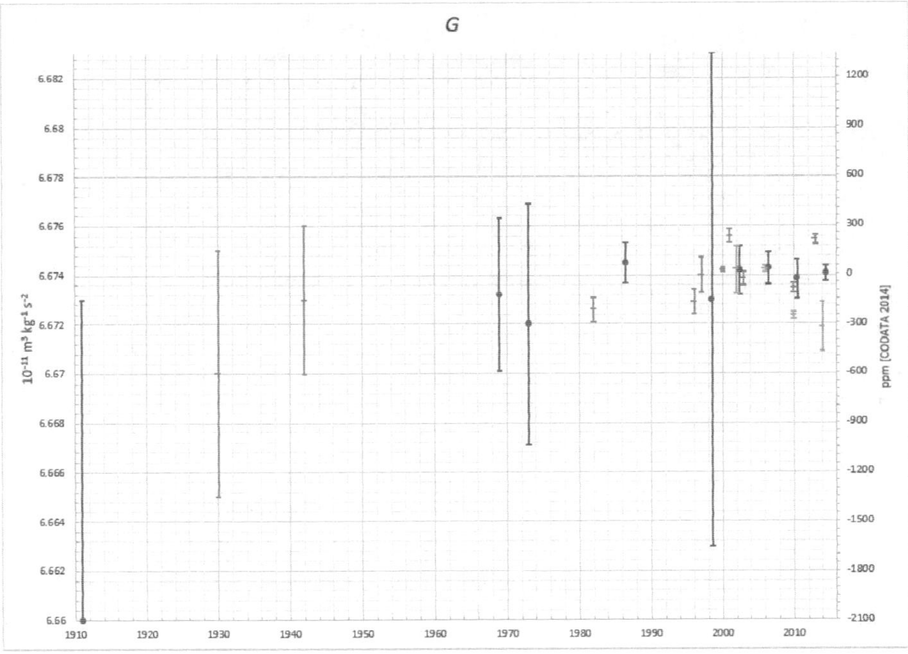

Figure 1. The results of measurements of the Newton constant G in different places on the Earth surface. The clear discrepancy between high-precision measurements during the last decades is still a serious mystery.

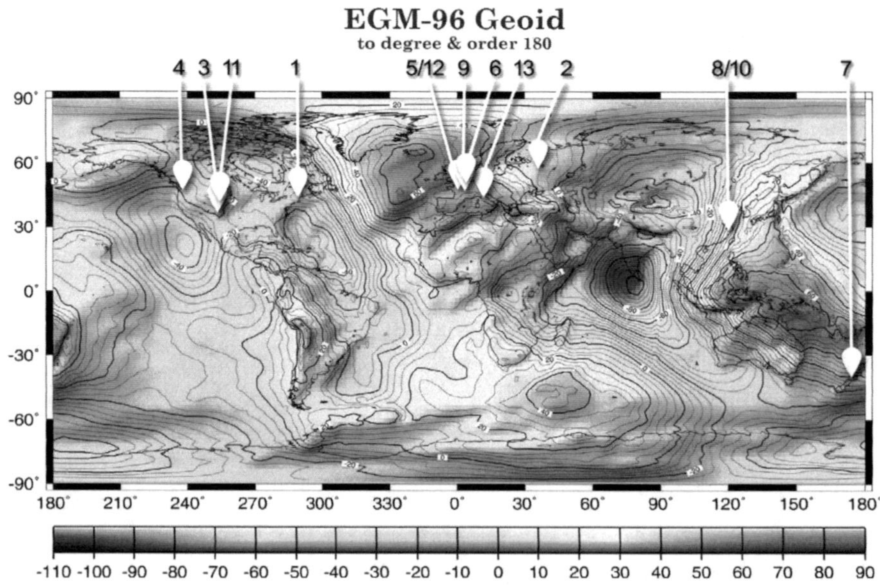

Figure 2. The pointers show the places where the data from Table 1 were obtained. As we see, these places are in a very narrow band on the Earth surface and do not give a representable picture of the values of Newton constant around the globe.

Table 1. The numerical results of the known high-precision measurements of the Newton constant G in units $G \times 10^{11}\,\mathrm{m^3\,kg^{-1}\,s^{-2}}$.

No	*G*	*Uncertainty*	*Altitude*	*Geoid*	*Latitude*	*Longitude*
1	6.67248(43)	6.4×10^{-5}	134.976	-30	39.13	-77.22
2	6.6729(5)	7.5×10^{-5}	151.785	10	55.76	37.62
3	6.6739(70)	1.0×10^{-4}	2160.594	-20	35.84	-106.29
4	6.674255(92)	1.4×10^{-5}	28.618	-30	47.66	-122.30
5	6.67559(27)	1.4×10^{-5}	89.050	50	48.82	2.21
6	6.67422(98)	1.5×10^{-4}	259.822	50	51.25	7.15
7	6.67387(27)	1.5×10^{-4}	0.000	20	-41.21	174.91
8	6.67228(87)	1.3×10^{-4}	38.477	10	30.51	114.41
9	6.67425(12)	1.9×10^{-5}	447.939	50	47.37	8.55
10	6.67349(18)	2.7×10^{-5}	38.477	10	30.51	114.41
11	6.67234(14)	2.1×10^{-5}	134.976	-20	40.01	-105.27
12	6.67545(18)	2.7×10^{-5}	89.050	50	48.82	2.21
13	6.67191(99)	1.48×10^{-4}	50.875	50	43.78	11.26
14	6.674484	1.161×10^{-5}	–	–	–	–
15	6.674184	1.164×10^{-5}	–	–	–	–

For search of time-variations of gravitational constant G see the recent paper [3] and the references therein.

Similar is the situation with the cosmological constant Λ at different places in the Universe [6], see Fig 3:

The main unsolved question: "Is the cosmological constant Λ really a constant, or it is some kind of field?", may be solved experimentally by proper measurements using high precision QSG

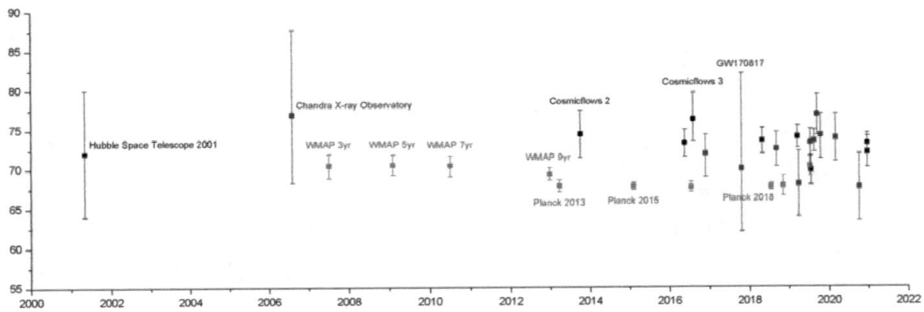

Figure 3. The results of measurements of the cosmological constant Λ in different places in the Universe. The observed in the last years discrepancy is also known as H_0-problem, where $H = \sqrt{\frac{\Lambda c^2}{3\Omega_\Lambda}}$ is the Hubble constant.

on satellites with large-altitude orbits and looking for post-Newtonian exponential tails in the gravitational potential V_{grav}:

$$V_{grav} = -\frac{Gm}{r}\left(1 - a_{vector}e^{-r/\lambda_{vector}} + a_{scalar}e^{-r/\lambda_{scalar}}\right).$$

For a recent review on this topic see [7, 8] and the references therein.

A more general and detailed consideration of the variation of constants can be found in [9].

The origin of the above experimental discrepancies is completely unclear.

It is not excluded that it lies in the properties, e.g. statics and dynamics of Dark Matter and Dark Energy, including their influence on the interior of the Earth, the Moon, the Solar system planets, as well as the Sun. Such influence is currently not sufficiently studied.

5. Absence of a more precise model than PREM [10] of the Earth interior as well as precise models of the interior of the Moon, the planets and the Sun.

6. The obvious conclusion is that today we know much more about the surrounding space than about the interiors of the Earth, the planets, their satellites and the Sun.

Very recently, some progress has been made in describing the Mars' interior [11].

2. Theoretical models

As a result of the circumstances, described in the points 1-6 of the Introduction, during the last several dozens of years a lot of different kind of modifications and extensions of GR were proposed and partially studied in the existing large amount of literature.

All these models cannot be considered to be true physical models of gravity. Instead, they present some mathematical speculations, which aim at the search of a future theory of gravity.

The main physical problem of these mathematical models is the absence of new physical principles for their construction.

In contrast, one can explain the vitality of GR by its deep and experimentally well-established physical principles on which GR was build.

Unfortunately, the basic GR principles are no longer sufficient for further development of the theory of gravity. For this purpose, novel physical principles are needed urgently. The most important issue is their experimental high-precision verification and confirmation.

Just as an example of extended GR model, we present here some preliminary results for possible DM and DE effects on the Earth's gravitational field, obtained in the framework of the simplest scalar-field-modification of GR, called Minimal Dilatonic Gravity (MDG) [12, 16, 17]. It is a further development and physical concretization of the O'Hanlon model of gravity [13].

In MDG, the Newton constant is replaced by an effective variable relativistic field $G_{eff}(\Phi)$, using a single scalar field Φ of Brans-Dicke type [14]. This field can be related with the known DM-effects. It is introduced according to the relations $G_{eff}(\Phi) = G/\Phi = G\,g(\phi)$.

In addition, the cosmological constant Λ, which is related with DE, together with a cosmological potential $U(\Phi)$, is introduced in MDG [1].

The action of the model is

$$\mathcal{A}_{g_{\mu\nu},\Phi} = \frac{c^3}{16\pi G}\int d^4x\sqrt{|\det(g_{\mu\nu})|}\,(\Phi R - 2\Lambda U(\Phi))$$

Just to illustrate the novel effects of DM and DE in the framework of MDG we can consider a simple model of the Earth, as a spherically symmetric non-rotating body with mass distribution, given by the experimentally established PREM [10].

This physical approximation is good-enough for qualitative consideration of the MDG-model of the Earth. For the first time, such model was applied to the gravitational field of the Earth in [2].

The hydrostatic equations of the static spherically symmetric bodies in MDG are a nontrivial generalization of the well-known Tolman-Oppehneimer-Volkov equations in GR, which, in its turn, generalize the hydrostatic Newton equations for static spherically symmetric objects, see for example, [15] and the references therein.

The detailed derivation of hydrostatic equations for four unknown functions $m(r)$, $\Phi(r)$, $p_\Phi(r)$, and $p(r)$ of the static spherically symmetric bodies in MDG:

$$\frac{dm}{dr} = 4\pi r^2\epsilon_{eff}/\Phi, \tag{1}$$

$$\frac{d\Phi}{dr} = -4\pi r^2 p_\Phi/\Delta, \tag{2}$$

$$\frac{dp_\Phi}{dr} = -\frac{p_\Phi}{r\Delta}\left(3r - 7m - \frac{2}{3}\Lambda r^3 + 4\pi r^3\epsilon_{eff}/\Phi\right) - \frac{2}{r}\epsilon_\Phi, \tag{3}$$

$$\frac{dp}{dr} = -\frac{p+\epsilon}{r}\frac{m + 4\pi r^3 p_{eff}/\Phi}{\Delta - 2\pi r^3 p_\Phi/\Phi}, \tag{4}$$

together with corresponding boundary conditions at the center of the Earth $r = 0$:

$$m(0) = m_c = 0, \quad \Phi(0) = \Phi_c, \quad p(0) = p_c,$$
$$p_\Phi(0) = p_{\Phi c} = \frac{2}{3}\left(\frac{\epsilon(p_c)}{3} - p_c\right) - \frac{\Lambda}{12\pi}V'(\Phi_c), \tag{5}$$

at the surface of the Earth with mean radius $r_{Earth} \approx 6371.0$ km:

$$m_{Earth} = m(r_{Earth}; p_c, \Phi_c), \quad \Phi_{Eurth} = \Phi(r_{Earth}; p_c, \Phi_c), \quad p_{\Phi_{Earth}} = p_\Phi(r_{Earth}; p_c, \Phi_c). \tag{6}$$

and at the cosmological horizon in the emerging de-Sitter Universe with huge unknown radius $r_{Universe} \sim 1/\sqrt{\Lambda} \approx 10^{22}$ km (in presence of only the Earth in it – as a crude approximation to the real Universe with huge total mass $M_{Universe} \gg m_{Earth}$):

$$\Delta(r_{Universe}; p_c, \Phi_c) = 0, \quad \Phi(r_{Universe}; p_c, \Phi_c) = 1 \tag{7}$$

[1] DM can be related either with some kind if new matter, outside the Standard Model of Particles (SMP), or by some modification of the theory of gravity in the style of MDG. The existence of DM is now well-established observationally in cosmology and astrophysics. Despite the intensive efforts and significant financial support, new massive matter particles, which spread with small velocities in comparison with velocity of light, were not found during the last decades, using the methods of particle physics.

Being also well established observationally in cosmology and astrophysics, DE is related with the very small cosmological constant $\Lambda \approx 10^{-44} km^{-2}$. From field-theoretical point of view, DE must constitute from particles, moving with the light velocity.

of the very nontrivial moving-boundary problem which appears in presence of a positive cosmological constant $\Lambda > 0$.

The last relation in Eq. (5) is a specific additional MDG condition of type $F_\Phi(p_{\Phi c} p_c, \Phi_c) = 0$ which ensures the finiteness of the DM pressure p_Φ at the center of the Earth.

The elimination of the unknown quantity $r_{Universe}$ from Eqs.(7) leads to second MDG-centre-values-relation in the form $F_\Lambda(p_c, \Phi c) = 0$. This procedure must be done numerically with high precision.

The dilatonic potential $V(\Phi)$ is defined by its derivative with respect to dilaton field Φ by the relation $V' = \frac{2}{3}\left(\Phi U'(\Phi) - 2U(\Phi)\right)$.

The general physical conditions on the cosmological potential $U(\Phi)$ are described in detail in [16]. Here we use the simplest cosmological potential $U(\Phi) = \Phi^2 + \frac{3}{16\Pi^2}\left(\Phi - 1/\Phi\right)$. In MDG. It plays the role of the harmonic oscillator of classical mechanics. This cosmological potential has a single physical vacuum of dilaton field, being a good approximation for any other cosmological potential $U(\Phi)$ in vicinity of its physical vacuum.

The parameter Π is the dimensionless Compton length (measured in cosmological units) of the dilaton and may have an extremely small values: $\Pi = \sqrt{\Lambda}\hbar/(m_\Phi c) \leq 10^{-30}$, due to the smallness of cosmological constant Λ and Planck constant $\hbar$.

There is no matter outside the Earth in above MDG model. Hence, there $p \equiv 0$ and $\epsilon \equiv 0$. In this domain we have a specific pure dilaton-sphere. Its structure is determined by the shortened system of differential equations in which the Eq.(4) has to be omitted.

In the above formulas (1) - (7), written down in so called geometrical units $c = 1, G = 1$ for simplicity, we have $\Delta = r - 2m - \frac{1}{3}\Lambda r^3$, and the very important for understanding of present article expression for the effective energy density and effective pressure

$$\epsilon_{eff} = \epsilon + \epsilon_\Lambda + \epsilon_\Phi, \quad p_{eff} = p + p_\Lambda + p_\Phi. \tag{8}$$

Here the energy density of the DE ϵ_Λ and the pressure p_Λ of the DE are

$$\epsilon_\Lambda = \frac{\Lambda}{8\pi}\left(U(\Phi) - \Phi\right), \quad p_\Lambda = -\frac{\Lambda}{8\pi}\left(U(\Phi) - \frac{1}{3}\Phi\right), \tag{9}$$

and obey the following specific MDG-equation-of-state of DE

$$\epsilon_\Lambda + p_\Lambda + \frac{\Lambda}{12\pi}\Phi = 0. \tag{10}$$

It explains very well the well-known empirical result of observational cosmology $\epsilon_\Lambda + p_\Lambda \approx 0$, because of the very small observed value of the cosmological constant Λ.

Using the area $A = 4\pi r^2$ of the surrounding the center of the Earth sphere of radius r, and the true infinitesimal geometrical distance $dl = dr\Big/\sqrt{1 - \frac{2m}{r} - \frac{1}{3}\Lambda r^2}$ in the three dimensional curved space, in geometrical units we obtain compact expressions for DM energy density and DM pressure:

$$\epsilon_\Phi = \frac{1}{8\pi}\frac{1}{A}\frac{d}{dl}\left(A\frac{d\Phi}{dl}\right), \quad p_\Phi = \frac{1}{8\pi}\frac{1}{A}\frac{dA}{dl}\frac{d\Phi}{dl}. \tag{11}$$

As it should be expected, these quantities do not contain explicitly the cosmological constant Λ. Instead, they are related to the gradients of the effective gravitational constant $G_{eff}(\Phi) = 1/\Phi$.

The MDG-equation-of-state of DM is much more complicated:

$$\epsilon_\Phi = p - \frac{1}{3}\epsilon + \frac{\Lambda}{8\pi}V'(\Phi) + \frac{p_\Phi}{2}\frac{m + 4\pi r^3 p_{eff}/\Phi}{\Delta - 2\pi r^3 p_\Phi/\Phi}. \tag{12}$$

It strongly depends on the equation of state of the usual mater of the Earth $\epsilon = \epsilon(p)$ which relates energy density ϵ and pressure p. See also the nontrivial relation between Figs. 4 and 7.

We actually do not need to know the explicit form of the equation of state $\epsilon = \epsilon(p)$. Indeed, to solve the problem, we can use in Eqs. (1)-(4) directly the PREM empirical data for the matter density $\rho(r)$ of the Earth, presented in Fig.4. Note that in geometrical units $\epsilon(r) = \rho(r)$. This is an essential simplification in our calculations.

Thus, in MDG one can conclude that the empirical knowledge of the function $\rho(r)$ from PREM is equivalent to knowledge of the complicated matter equation of state of the Earth.

Now, we have a well defined moving boundary problem for our model of the Earth structure in MDG.

The analytical and numerical analysis for neutron stars, which defers from the above MDG model of the Earth only by mater equation of state, can be found in [16, 17]. In realistic models of neutron stars, in the framework of this model, one has to assume the mass of the scalar field to be $m_\Phi c^2 \approx 4.3 \times 10^{-13}$ eV [17, 18]. This value of $m_\Phi c^2$, gives for Compton length of the dilaton $\lambda_\Phi \approx 3000$ km. Here we adopt this value of the Compton length of the dilaton in the MDG model of the Earth, assuming that in the two cases the dilaton is the same particle.

The preliminary results for the Earth are shown in the four figures 4 – 8 below:

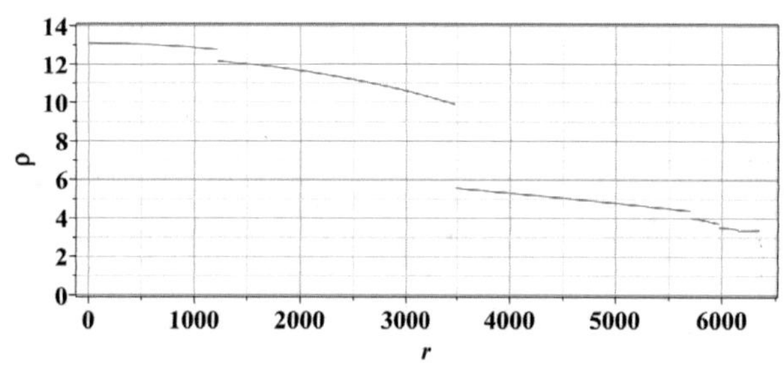

Figure 4. The density $\rho(r)$ in unites $\frac{g}{cm^3}$ of the matter inside the Earth according to the PREM [10]. The distance r from the Earth center is in units km

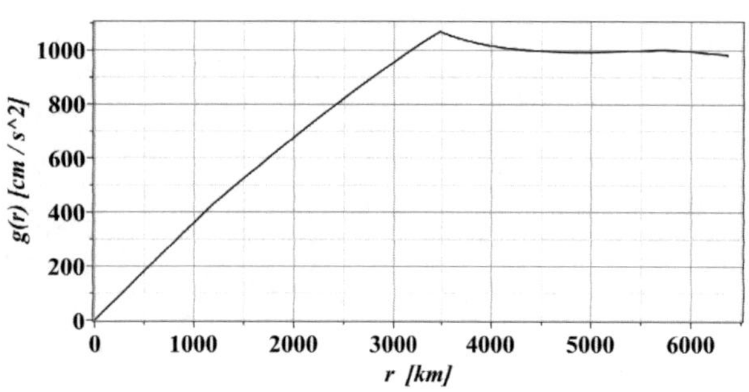

Figure 5. The PREM-gravitational-acceleration inside the Earth in Newton gravity.

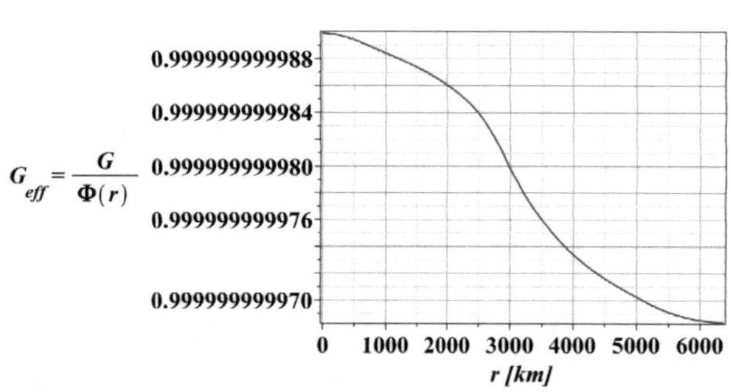

Figure 6. The effective field G_{eff} inside the Earth, which replaces the Newton constant G in the MDG model.

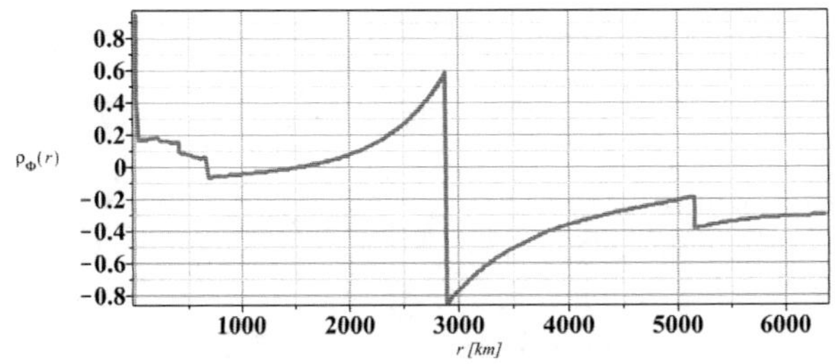

Figure 7. The MDG-Dark-Matter-profile inside the Earth. Note that it follows the PREM-matter-density-profile in Fig. 4 in some specific way.

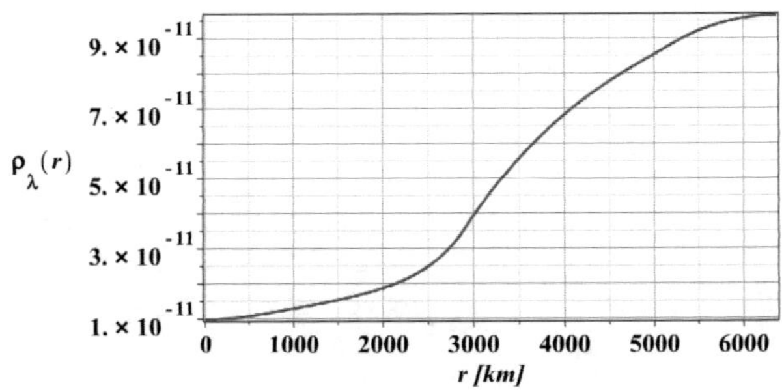

Figure 8. The MDG-Dark-Energy-profile inside the Earth. Note that it is completely different from MDG-Dark-Matter-profile in Fig. 7.

As seen in Figs. 4-8, in framework of the MDG model, the effective gravitational constant G_{eff} , as well as effects of DM and DE indeed depend on the place in three-dimensional space. In our crude MDG-model of the Earth we assume exact spherically symmetric distribution of the usual matter inside a non-rotating Earth and absence of other bodies in the Universe, outside the Earth. As a result, we obtain that G_{eff} , as well as DM and DE depend only on the distance to the center of the Earth.

It is obvious, that any inhomogeneities inside the Earth, as well as presence of other bodies around the Earth will destroy spherical symmetry. In this case, the dependence of G_{eff} on the place in three dimensional space, as well as DM and DE will be much more complicated. These dependencies may be similar to dependencies, shown in Figs. 1 and 2, and in Table 1, at least to some extend.

For now, our present consideration shows the real possibility to explain the empirical data by proper new model of gravity in the style of MDG-model. For this purpose two basic things are needed:

1. Experimental hi-precision data – to find the physical principles for building new physically valuable model of gravity, as extension of GR.

2. Much more complicated numerical analysis of the model of the Earth, taking into account the real inhomogeneities inside the Earth, its rotation, real form, hydrosphere and atmosphere, as well as surrounding the Earth space environment.

The solution of these two hard problems we leave for future investigation in collaboration with proper scientific organizations.

3. Conclusion

1. In the present-days real state of affairs, the interpretation of data obtained from high-precision measurements by satellites for geodetic use seems to be quite problematic, uncertain, and may be misleading for practitioners.

2. In physics we have no absolute truth and absolute theories.

All we can do, is to create physical models and compare them with the physical reality with the available precision at given time instant.

When the precision increases by 2-3 orders of magnitude, due to the technological progress, as a rule, we observe novel phenomena, previously unknown.

This explains the boundaries of applicability of any old model. As a result, the old model must be replaced with a new one.

3. Let me point out a known example, directly related with geodesy:

In the framework of measurements of distances with a relative precision of 10^{-4}, the geometry of space around the Earth is Euclidian one.

It was Gauss, who checked with this precision that the sum of the angles in a triangle is 180^0, using three mountain peaks around Geneva. His result supported the Euclidian geometry.

Therefore, Gauss stopped developing the geometry of curved space, leaving this issue for his PhD student Riemann.

Due to more recent developments of technology (Mössbouer effect) the measurements with precision 10^{-7} prove the non-Euclidean geometry of space around the Earth surface and supported GR [19, 20].

It turns out that the sum of the angles in a geodesic triangle on the Earth surface is larger than 180^0.

Note that for such basic observation on the surface of a neutron star, it is enough to have precision of 10^{-2}.

4. At small enough distances and velocities, the motion of elementary particles is described by QM, not by Classical Mechanics (CM). Quantitative measure for the new physics is given by a comparison of CM action with the Planck constant, which is very small from a macroscopic point of view. Such quantitative measure needs high-precision experiments.

Decisive condition for applicability of CM- or QM-description of assembles of particles in macroscopic scales (Bose condensate, assemble of cold atoms etc.), included in Quantum Space Gravimetry (QSG), is the absence or presence of coherent-macro-state. It is necessary that for its creation there are physical conditions realized with high precision.

5. There is no any theoretical basis that could guarantee that after increasing by 3-4 orders of magnitude of QSG precision for geodetic purposes, we still would be able to use the Newton model of gravity created more than three centuries ago, and based on linear equations (Laplace equation) and linear superposition principle (Poison equation).

Note that the Poison equation is a basis for analysis of data from satellites in geodesy since it is assumed that any deviations of the gravitational field of the Earth from the average one are caused by local space or time variations of the matter density in the Earth interior, in the oceans and seas, in the polar ice, in the Earth atmosphere, etc. [21].

This assumption cannot be true in nonlinear models of gravity like GR and its modifications. Without careful analysis, we cannot be sure that the routine approach to this problem correctly describes reality by the linear superposition principle of the Newton model.

Just the opposite, it is physically natural to expect that the high-precision observational data will not be correctly interpreted using the oldest Newton model of gravity, with the well-known limits of applicability.

6. Based on the Newton model of gravity is the interpretation of the modern satellite data in large amount of important real cases [22]:

- Ice-sheet loss at North Pole

- Glacier and ice-cup loss in North America
- Precipitation increase in North America
- Glacier retreating in West-North America
- Surface water drying in Central North America
- Progression from dry to wet period Central North America
- Groundwater depletion and drought West Cost of North America
- Drought in Texas
- Recovery from early-period drought North-South America
- Recent drought in Central-South America
- Patagonian ice-field melt
- Ice-sheet loss at South Pole
- Decline the Aral Sea
- Decline the Caspian Sea
- Precipitation increase in Asia
- Glacier melt, surface water in Asia
- Groundwater depletion in Cine
- Water depletion and precipitation decrease in India
- Groundwater depletion in South Asia and Iran
- Progression from dry to wet period in East Australia
- Return to normal after wet period in North Australia
- Groundwater depletion in North Africa
- Increasing lake levels and groundwater in East Africa
- Precipitation increase in the West Africa
- Precipitation decrease in the South Africa
- Precipitation increase in the South-East Africa

etc.

After achieving higher precision in several orders of magnitude, the new satellite data may be misleading for practitioners, because of the wrong interpretation based on Newton model of gravity and use of still standard in geodesy Euclidean geometry.

7. Let us mention some of the basic achievements of modern relativistic geodesy [23, 24], which have to be taken into account in high-precision satellite measurements:

- Time and Frequency Metrology in Context of Relativistic Geodesy
- Chronometric Geodesy
- Measuring Gravitational Field in GR
- GR Gravity Gradiometry
- Reference -Ellipsoid End Normal Gravity Field in Post-Newtonian Geodesy
- Use of Geodesy and Geophysics Measurements to Probe Gravitational Interactions
- Use of Space-Time Curvature in Future Navigation Systems
- Measurement of Frame Drugging with Geodetic Satellites (CHAMP, GRACE and beyond)
- Test of GE with LARES Satellites

We stress that the problem of the nonlinear principle of superposition in GR and its modification is a completely open issue. Some initial attempts to formulate this problem were performed in [25].

8. All the above arguments require deep analysis and preliminary study of the real situation with high-precision QSG for geodetic purposes, both theoretically and experimentally.

We definitely need to try to avoid possible serous mistakes when executing expensive projects like QSG Pathfinder under QSG Program of EC and ESA.

In my opinion, for this purpose, it is necessary to perform deep preliminary theoretical and experimental studies and to use existing scientific organizations and structures at national and international level, proper scientific collaboration in the framework of EU as well as a wider international collaboration in the framework of ESA.

A recent survey on quantum physics in space, including some of the above issues, different modern achievements, open problems, perspectives, existing different approaches and results can be found in the very recent and more detailed review [26, 27].

Acknowledgments

The author is deeply indebted to the Directorate of the Laboratory of Theoretical Physics, JINR, Dubna, for the good working conditions and support during my work there in 2013-2021.

This research was supported in part by the Foundation for Theoretical and Computational Physics and Astrophysics and by Bulgarian Nuclear Regulatory Agency, Grant for 2021.

The author is also tankful to the organizers of the Quantum Space Gravimetry (QSG) for Earth Observation Round Table, Brussels, 14 April, 2021; COM-ESA Workshop on the user requirements for Space (Quantum) Gravimetry , Brussels, 30 September –1 October, 2021; Conference NAFSKI-2021, Sofia, 07-09 October 2021; and Quantum Space Gravimetry - Technical Configuration Nr.2 of the Consultation Platform, Brussels, 12 November 2021; for possibility to give talks and discuss the basic ideas and results of the present paper.

Special tanks to the unknown referee for the useful suggestions and remarks.

References

[1] Gravitational constant - Wikipedia

[2] Fiziev P P, Fizieva Ts Ya, 2014, Variability of the Gravitational Constant - Forty-Three Years Later, *Conference Fundamental Physical Constants* **14**, in Conference proceedings

[3] S. Bhagvati, S. Desai, 2021, Search for variability in Newton's constant using local gravitational acceleration measurements, arXiv:2108.05012

[4] G. Rosi, F. Sorrentino, L. Cacciapuoti, M. Prevedelli, G. M. Tino, 2014, Precision measurement of the Newtonian gravitational constant using cold atoms, NATURE, Vol. 510, 518

[5] Absolute Quantum Gravimeter (A free-fall absolute gravimeter based on laser-cooled atoms), https://www.muquans.com/

[6] Hubble's law - Wikipedia

[7] B. Buscaino, D. DeBra, P. W. Graham, G. Gratta, T. D. Wiser, 2015, Testing long-distance modifications of gravity to 100 astronomical units, Phys. Rev. D 92, 104048

[8] Y.-J. Chen, W. K. Tham, D. E. Krause, D. López, 5 E. Fischbach, R. S. Decca, 2016, Stronger Limits on Hypothetical Yukawa Interactions in the 30–8000 nm Range, PRL **116** 221102

[9] Jean-Philippe Uzan, 2011, Varying Constants, Gravitation and Cosmology, Living Rev. Relativity, 14, 2; https://link.springer.com/article/10.12942/lrr-2011-2

[10] Dziewonski A M, Anderson D L, 1981, Preliminary reference Earth model, *Physics of the Earth and Planetary Interiors,* **25** 297

[11] Martin van Driel et. all., 2021 High-Frequency Seismic Events on Mars Observed by InSight, *JGR: Planets,* **126** 2

[12] Fiziev P P, 2000, A minimal model for dilatonic gravity *Mod Phys Lett A,* **15** 32

[13] O'Hanlon, 1972, Phys. Rev. Lett. **29** 137.

[14] C. Brans, R. H. Dicke, Phys. Rev. **124** 925 (1961).

[15] S. L. Shapiro, S. A. Teukolsky, 1983. Black Holes, White Dwarfs and Neutron Stars, A Wiley-Interscience Publication, Cornel University, Itaca, N.Y.

[16] Fiziev P P, 2013, Withholding potentials, absence of ghosts, and relationship between minimal dilatonic gravity and $f(R)$ theories, *Phys. Rev. D* **87** 044053

[17] Fiziev P, Marinov K, 2017, Modeling of nonrotating neutron stars in minimal dilatonic gravity, *Astrophys Space Sci,* **362** 8

[18] Fiziev P, 2017, *Mod Phys Lett A* **32** 1750141

[19] Pound R V, Rebka G A Jr, 1959, Gravitational red-shift in nuclear resonance, *PRL* **3**

[20] Pound R V, Snider J L,1964, Effect of gravity on nuclear resonance *PRL* **13** 2

[21] Geophysical Surveying Using Gravity https://www.ukm.my/rahim/G-Gravity

[22] Rodell et al., 2018, Emerging trends in global freshwater availability. *Nature* **557** *651*

[23] Bjerhammar A, Relativistic Geodesy, NOAA Technical Report NOS 1 18 NGS 36 (1986)

[24] Relativistic Geodesy, Foundations and Applications, 2019, Editors Dirk Puetzfeld, Claus Lämmerzahl, *Fundamental Theories in Physics,* **196** *Springer Nature Switzerland AG*
[25] Fiziev P P, 2007, Static Fundamental Solutions of Einstein Equations and Superposition, Preprint arXiv:gr-qc/0701108
[26] Belenchia A et al, 2021, Quantum Physics in Space, Preprint arXiv:2108.01435
[27] Iv´an Alonso et al, 2022, Cold Atoms in Space: Community Workshop Summary and Proposed Road-Map, arXiv:2201.07789

Offline noise calibration of the CAEN V1742 ADCs at the PADME experiment

Georgi Georgiev, for the PADME collaboration [1]
Institute for Nuclear Research and Nuclear Eenergy, Bulgarian Academy of Sciences, BG
Physics Faculty, Sofia University "St. Kliment Ohridski", BG

E-mail: georgiev.georgi.st@gmail.com

Abstract. This work aims to provide robust methods for offline calibration of the noise floor for CAEN's V1742 sampling ADCs, used at the PADME experiment. It has been found that a part of the noise in the final signals originates from not fully equalized storage capacitor properties in the DRS4 chip, even after application of the CAEN's proprietary *drs4* calibration. Two approaches to offline equalization are discussed and finally two methods are proposed for obtaining the calibration data from random and physics trigger runs.

1. Introduction

The PADME experiment is dedicated to dark photon searches in the process of annihilation of accelerated positrons with electrons in a fixed target [1, 2]. For the digitization of $\mathcal{O}(1000)$ of analog channels at the experiment 29 CAEN V1742 ADC boards are in use. A custom made trigger distribution system is in charge of synchronizing all the 29 boards [3]. The design of the data acquisition system (DAQ) allows for complete offline analysis with no prior knowledge for the signals, as the entire ADC waveforms are stored.

The ADC board V1742 is equipped with 32 analog input channels, 2 trigger input channels and several extra interfaces: an optical connection for data transmission, common trigger I/Os that can handle differential or NIM signals and a number of extra general purpose LVDS inputs/outputs. The key element of the digitizer board is the DRS4 chip, which was developed at the Paul Scherrer Institut for the MEG experiment [4]. The DRS4 is a low noise switching capacitor buffer, capable of sampling 9 channels with 1024 storage capacitors for each channel.

[1] A.P. Caricato, M. Martino, I. Oceano, F. Oliva, S. Spagnolo (INFN Lecce and Dip. di Matematica e Fisica), G. Chiodini (INFN Lecce), F. Bossi, B. Buonomo, R. De Sangro, C. Di Giulio, D. Domenici, G. Finocchiaro, L.G. Foggetta, M. Garattini, A. Ghigo, F. Giacchino, P. Gianotti, I. Sarra, B. Sciascia, T. Spadaro, E. Spiriti, C. Taruggi, E. Vilucchi (INFN Laboratori Nazionali di Frascati), V. Kozhuharov (University of Sofia "St. Kl. Ohridski" and INFN Laboratori Nazionali di Frascati), G. Georgiev (Institute for Nuclear Research and Nuclear Eenergy, Bulgarian Academy of Sciences and University of Sofia "St. Kl. Ohridski"), S. Ivanov, R. Simeonov (University of Sofia "St. Kl. Ohridski"), F. Ferrarotto, E. Leonardi, F. Safai Tehrani, P. Valente (INFN Roma1), E. Long, G.C. Organtini, G. Piperno, M. Raggi (INFN Roma1 and Dip. di Fisica, "Sapienza" Università di Roma), S. Fiore (ENEA centro ricerche Frascati and INFN Roma1), V. Capirossi, F. Iazzi, F. Pinna (Politecnico di Torino), B. Liberti (Dip. di Fisica, Università di Roma "Tor Vergata", INFN Laboratori Nazionali di Frascati), M. Martini (Università G. Marconi and INFN Laboratori Nazionali di Frascati), J. Alexander, A. Frankenthal (Dep. of *Physics* Cornell University).

The sampling rate of the DRS4 chip is limited to 5 GS/s. A noise level of 0.35 mV is achievable after an individual capacitor-offset calibration [5].

There is a drs4 calibration procedure implemented in the digitizer firmware, that is configurable through the CAENDigitizer library interface [5]. Although the impact of this procedure is very large, we found that there is room for an additional offline calibration. A further data correction can be applied to the data collected with the stock calibration table, loaded in the hardware. Even if it is clearly stated that DRS4 has a radiation hard design [5], the proposed method can also correct for hardware degradation of the electronics.

The majority of the detectors in the PADME experiment are scintillating detectors [3] and produce significant signals. So the effect of an improvement in the ADC noise level is expected to provide a relatively small impact on the overall energy resolution. On the other hand, the proposed calibration is expected to lead to better time resolution and improved double pulse separation.

2. Methods and Materials

For the analysis, two data sets of 2020 run were used, later in text referred to as *Physics* and *Random* runs, respectively. During the *Physics* run, all detectors were operational, a beam with energy 440 MeV and intensity 25 kPOT was delivered to the experiment, and the trigger signal was provided by the accelerator. During the *Random* run no beam was delivered, but the trigger source remained the same. This special condition is equivalent to a random trigger which makes this particular run suitable for noise floor studies. Only the first 5000 events were used for both runs.

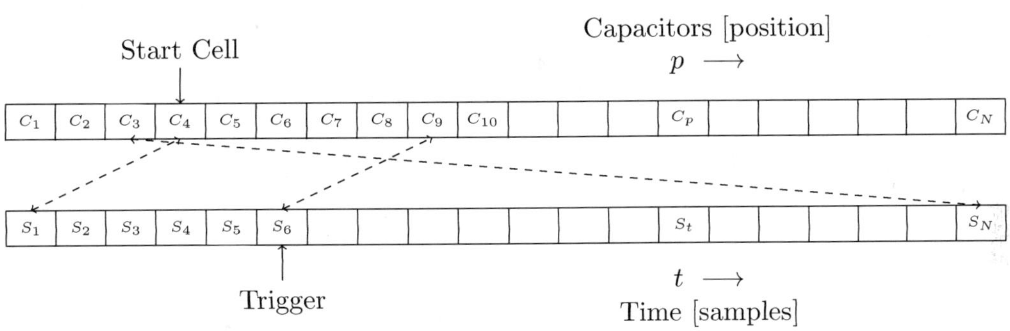

Figure 1. Illustration of the relationship between capacitors and time positions. The lower row represents samples S in time, while the upper one represents the capacitors in the circular buffer.

The data are organized in events. Each event contains information about the enabled ADCs, and their channels. 1024 amplitude samples are available per channel together with information about the starting position of the DRS4 circular buffer – *StartCell*. It is visible from figure 1 that the relationship between the time t and the capacitor position p is:

$$p = (t + StartCell) \bmod 1024 \tag{1}$$

As the triggers start the digitization process, they will appear almost at the same time on the waveform. In different events, the *StartCell* may change leading to a different correspondence between the capacitor position in the circular buffer p and time step t. In order to calibrate the effects of inequalities of the capacitors it is not enough to subtract the pedestal for each time slot as multiple capacitors could be associated at a particular time slot. In accordance to the described model the calibration looks like:

$$S_{cor}(t) = S_{meas}(t) - E_{p(t)}(t), \tag{2}$$

where S_{meas} is the measured amplitude for a given time t, S_{cor} is its corrected value, $E_{p(t)}$ is the calibration function for the capacitor at position p at time t. Note the dependence on p of t (equation 1).

Different methods can be defined to obtain the calibration function $E_{p(t)}(t)$. Often, when access to raw data is needed, the researchers will need to run the reconstruction by their own, so the chosen strategy should satisfy these requirements: (i) the result should be obtained after a single pass over the data; (ii) the required computer resources should be compatible with a nowadays average computer; (iii) the method in any case should not worsen the noise condition.

In order to process the data on capacitor basis, the perspective should change from sequence of amplitudes ordered in time to sequence of amplitudes ordered by capacitor position using equation 1. Two main approaches were studied.

The first approach takes the mode of the amplitude distribution for each particular capacitor as a constant estimate of E_p. The method is realized by collecting per-capacitor histograms of the amplitudes for all the events in the data set. Once all events are processed, the value of the mode of the distribution can be determined.

The second approach accounts not only for the capacitor offsets but also for the time-dependent capacitor characteristics, for example parasitic currents charging or draining the storage capacitors. It relies on performing polynomial fits on amplitude distribution over time for each capacitor. Collecting per-capacitor 2D histograms for the amplitude values over time is not an option as it violates (ii) from the requirements list. In the present approach, the parameter's estimation is done by performing a χ^2 minimization procedure, based on the collection of a set of sums [6]:

$$a_{p,0} \cdots a_{p,k} = \arg \min_{a_{p,0} \cdots a_{p,k}} \underbrace{\sum_{events} (S_p(t) - E_p(t))^2}_{\chi^2}, \tag{3}$$

where $a_{p,0} \cdots a_{p,k}$ are the polynomial parameters for capacitor p, t is the time position where the capacitor at position p appears to be for the current event, k is the degree of the polynomial and

$$E_p(t) = \sum_{i=0}^{k} a_{p,i}\, S(t)^k. \tag{4}$$

In order to calculate the polynomial parameters the following sums should be collected:

$$\Sigma_{x^i} = \sum_{events} t^i, \quad \forall i \in [0, 2\,k] \tag{5}$$

$$\Sigma_{x^i y} = \sum_{events} t^i\, S(t), \quad \forall i \in [0, k]. \tag{6}$$

For polynomials up to 5-th degree the total number of sums is 17: $\sum x^0, \cdots, \sum x^{10}$ and $\sum x^0 y, \cdots, \sum x^5 y$.

The source of floating point roundoff errors is eliminated by using integer variables for collecting the sums. By using 16 byte variables more than 10×10^8 events can be processed without overflow. This is well within the recommended limit for the number of events in a single run at the PADME experiment.

Both the statistical mode of the first approach and the a_0 of the polynomial approach are measures for the pedestal of the signals. In order to apply the corrections by preserving the initial offset, corrected offsets o' are to be calculated as:

$$o'_i = o_i - \frac{\sum_{j=1}^{1024} o_j}{1024}. \tag{7}$$

3. Results and Discussion

The time dependence of the amplitude for a single capacitor is shown on figure 2 as violin plots. The widths of the violins represent the occupancy at a given amplitude at a given time. Amplitude spectra are shown on right side of the figures. Figure 2a shows data from the *Random* run, while figure 2b shows data from the *Physics* run. These results are obtained in the same manner – the studied capacitor is №16 of channel 12 of ADC 27. This channel has been selected as it corresponds to a central crystal of the small angle electromagnetic calorimeter of PADME, that is exposed to the highest particle rate, so the effects of signal presence will be more pronounced.

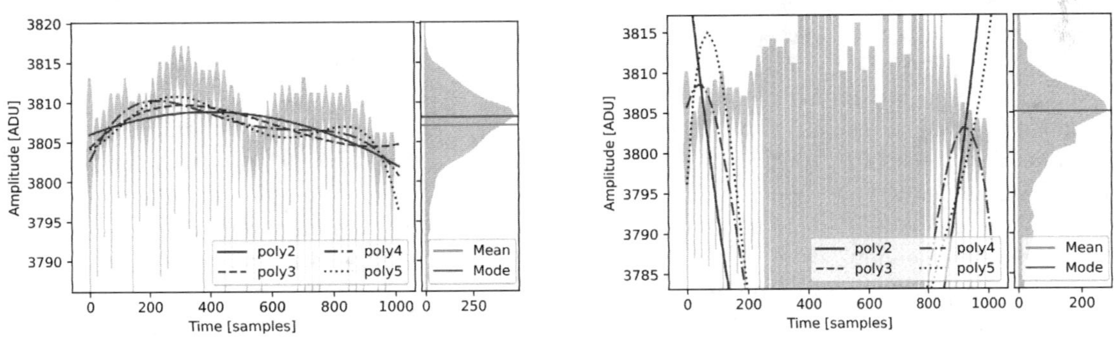

(a) *Random* run (b) *Physics* run

Figure 2. 5000 events ADC 27, channel 12, capacitor 16. The width of the violins represent the population at a given amplitude.

In figure 2a a reoccurring pattern is noticeable, which is an indication that such a calibration is applicable. Also it is visible that polynomials with higher degree approximate the data better. On the contrary figure 2b has a region with many signals starting at about $t = 250$ sample. All polynomials fit to the signals and failed to provide a measure for the noise floor. The mode of the amplitude distribution remained a good estimate of the offset of the capacitor.

Three combinations were tested:

(a) *Random* run with *Random* run calibration
(b) *Physics* run with *Random* run calibration
(c) *Physics* run with *Physics* run calibration

Results for (a) and (b) are provided in figure 3, where each violin shows the RMS of the noise distribution for all the channels with a given algorithm. Results for (c) are not shown on the figure as they copy the result for the mode in (b) and the results for polynomial calibration was worse than initial.

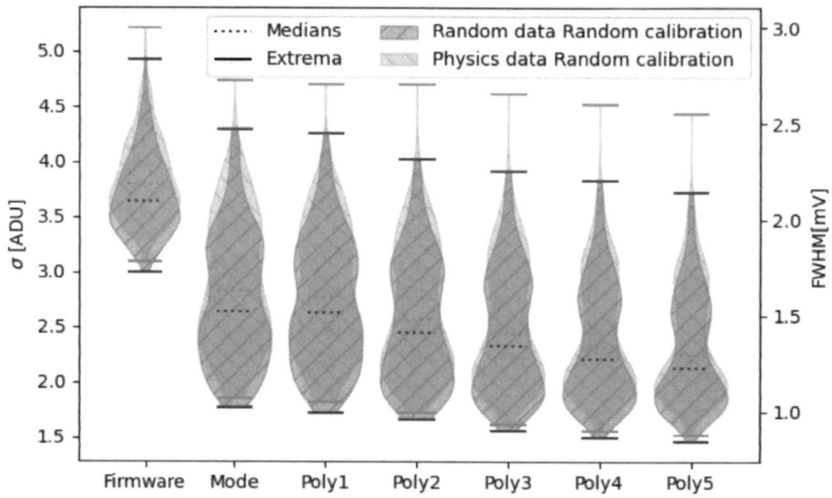

Figure 3. Comparison between different calibration methods. To obtain this plot 5000 events were used from both *Random* and *Physics* runs. Channels 16-24 of ADC27 are excluded from the statistics as they are known to be noisier.

The fact that the results from (a) and (b) are very close means that the pattern of capacitor inequalities is preserved over the time and it is possible to use calibration parameters obtained from different runs.

The structure in the shape of the violins corresponding to higher degree polynomials mean that there are other sources of noise which are individual to each particular channel and became dominant. So it will be very difficult to do any further improvement using these methods. The application of the algorithms shows significant improvement. A nearly two times better noise condition is achieved – starting at about 2 mV and getting lower than 1.2 mV in the case of Poly5 algorithm.

The suggestion is to use runs with *Cosmic* or *Random* triggers (or other runs with low counting rates) to calculate the calibration parameters with Poly5 method which can be used to calibrate *Physics* runs afterwards. If only runs with high counting rates are available, it is suggested to use the Mode method.

4. Conclusions

It has been found that CAEN's proprietary calibration, implemented in the firmware, does not produce the best possible noise condition and it can be further improved by running an offline calibration algorithm. Two main approaches to improve noise condition in the V1742 digitizers at PADME experiment were proposed and discussed. Such an improvement may lead to better time resolutions or easier peak separation.

Acknowledgments

This work is supported by MUCCA, CHIST-ERA-19-XAI-009 / BG-NSF KP-06-D002/4 from 15.12.2020 and funded by The Bulgarian national program "Young scientists and postdocs" RMS 577/17.08.2018 and by Sofia University, Bulgaria.

References

[1] Raggi M., Kozhuharov V., Valente P. The PADME experiment at LNF, *EPJ Web of Conferences* (2015) **96**.
[2] Raggi M, Kozhuharov V, Proposal to Search for a Dark Photon in Positron on Target Collisions at DAΦNE Linac", *Advances in High Energy Physics*, vol. **2014**, Article ID 959802, 14 pages, 2014.
[3] E Leonardi et al. 2017 *J. Phys.*: Conf. Ser. **898** 032024.
[4] V1742 & VX1742 32+2 Channel 12bit 5 GS/s Switched Capacitor Digitizer, 2020, *CAEN SpA* rev. 8.
[5] 9 Channel, 5 GSPS Switched Capacitor Array DRS4, *Paul Scherrer Institut*, Rev. 0.9.
[6] Press, William H.; Teukolsky, Saul A.; Vetterling, William T.; Flannery, Brian P., Numerical Recipes 3rd Edition: The Art of Scientific Computing, December 2007, *Cambridge University Press.*

PADME physics program

Radoslav Simeonov [1]

Faculty of Physics, University of Sofia "St. Kl. Ohridski",
5 James Bourchier Boulevard, BG-1164 Sofia, Bulgaria

E-mail: radoslav.simeonov@cern.ch

Abstract. The complete nature of the dark matter stays unsolved in the terms of the Standard Model. A possible solution suggests complete new class of particles which participate in interactions with Standard Model particles via newly introduced mediators, the so called portals. The Positron Annihilation into Dark Matter Experiment (PADME) aims to observe indirectly one of these portals – the dark photon A', through the reaction $e^+ + e^- \to \gamma + A'$. Two data acquisition runs were already performed in order to study detectors behavior, background levels and mainly to throw light in the A' mass region for 2 - 23.7 MeV/c^2.

1. Introduction

The particles from the Standard Model (SM) can not consistently describe the phenomenon of dark matter (DM). One of the possible solutions in contemporary high energy physics suggests the existence of a whole new dark sector of particles, some of which may act as DM constituents. These particles may interact with the SM extremely feebly, which is the main reason why it is hard to detect them. One of the simplest models adds a U(1) symmetry with a corresponding vector boson particle, acting as a portal between the two sectors – the dark photon A'. This boson may be massive and mix kinetically with the SM photon, parametrised by the mixing constant ε. In such scenario A' can be produced in lepton annihilation, bremsstrahlung and meson decays (where the mesons could also be produced in e^+e^- interactions). Two possibilities exist for the decay of the A':

[1] For the PADME collaboration: A.P. Caricato, M. Martino, I. Oceano, F. Oliva, S. Spagnolo (INFN Lecce and Dip. di Matematica e Fisica, Università del Salento Via Provinciale per Arnesano, 73100 Lecce) G. Chiodini (INFN Lecce Via Provinciale per Arnesano, 73100 Lecce) F. Bossi, B. Buonomo, R. De Sangro, C. Di Giulio, D. Domenici, G. Finocchiaro, L.G. Foggetta, M. Garattini, A. Ghigo, F. Giacchino, P. Gianotti, I. Sarra, B. Sciascia, T. Spadaro, E. Spiriti, C. Taruggi, E. Vilucchi (INFN Laboratori Nazionali di Frascati Via E. Fermi 54, 00044 Frascati) V. Kozhuharov (Faculty of Physics, University of Sofia and INFN Laboratori Nazionali di Frascati, 5. J Bourchier Blvd, 1164 Sofia, Bulgaria) G. Georgiev, S. Ivanov, R. Simeonov (Faculty of Physics, University of Sofia "St. Kl. Ohridski", 5. J Bourchier Blvd, 1164 Sofia, Bulgaria) F. Ferrarotto, E. Leonardi, F. Safai Tehrani, P. Valente (INFN Roma1 p.le Aldo Moro 5, 00185 Roma) E. Long, G.C. Organtini, G. Piperno, M. Raggi (INFN Roma1 and Dip. di Fisica, "Sapienza" Università di Roma p.le Aldo Moro 5, 00185 Roma) S. Fiore (ENEA centro ricerche Frascati and INFN Roma1 Via Enrico Fermi 45, 00044 Frascati (Roma)) V. Capirossi, F. Iazzi, F. Pinna (Politecnico di Torino, Corso Duca degli Abruzzi, 24, 10129 Torino) B. Liberti (Dip. di Fisica, Università di Roma "Tor Vergata", INFN Laboratori Nazionali di Frascati Via della Ricerca Scientifica 1, 00133 Roma) M. Martini (Università G. Marconi and INFN Laboratori Nazionali di Frascati Via Plinio 44, 00193 Roma) J. Alexander, A. Frankenthal (Dep. of Physics Cornell University, 109 Clark Hall, Ithaca,14853 New York)

- Supposing that A' is the lightest particle from the dark sector with mass $m_{A'} > 2m_e$, it undergoes visible decays into SM particles, e.g. $A' \rightarrow e^+e^-$.
- If lighter particles than the dark photon exist in the dark sector ($m_{A'} > 2m_{DM}$, where m_{DM} is the mass of any lighter than A' dark sector particle), then the dark photon may decay invisibly.

The need of only two parameters (ε and $m_{A'}$) to describe the dark photon justifies the high predictability and popularity of the model.

Several other possible solutions to the DM problem are also viable, namely Axion-Like Particles (ALPs), Dark higgs, weakly interacting massive particles, etc. The variety of models leads to a possible mass of the dark matter candidate spanning more than 15 orders of magnitude.

2. The PADME experiment

Positron Annihilation into Dark Matter Experiment (PADME) [1], located at the Beam Test Facility of the Laboratori Nazionali di Frascati (LNF) in Italy, aims to seek for a dark photon through the interaction $e^+ + e^- \rightarrow \gamma + A'$. It uses a positron beam, provided by the DAΦNE LINAC [2]. The beam can reach energy of 550 MeV. The experiment uses a missing mass technique to determine the mass of A'. The missing mass squared is given by

$$M^2_{miss} = (P_{e^+_{beam}} + P_{e^-} - P_\gamma)^2, \tag{1}$$

where $P_{e^+_{beam}}, P_{e^-}, P_\gamma$ are the 4-momenta of the positron, of the electron and of the emitted γ respectively. The A' will appear as a peak at a fixed position in the missing mass distribution. The maximum energy of the beam limits the sensitivity of the experiment up to $m_{A'} = \sqrt{2m_e E_{beam}} \simeq 23.7$ MeV/c^2.

2.1. Detector system

A schematic layout of the PADME detector system is given in fig. 1. It is a small scale fixed target experiment - the distance from the target to the electromagnetic calorimeter front face is less than 3.5 m.

The experiment uses an active target made by 100 µm thick polycrystalline diamond. It serves as annihilation media for the positrons, but it also acts as a detector for precise determination of the position and multiplicity of the incoming beam. The low Z of carbon and the thickness of the diamond target have been optimized to keep the background from the bremsstrahlung events[3] under control. Opposite to the diamond target a monolithic active-pixel sensor MIMOSA-28 is mounted, which is operated in vacuum. It is able to monitor the beam when the active target is out and could provide information for the beam divergence and position. After the interaction point, the beam enters a large vacuum chamber (pressure less than 10^{-6} mbar) with ~ 0.5T magnetic field, provided by a dipole magnet. Alongside the beamline inside the vacuum chamber two segmented charge particle detectors (vetoes) are placed. They consist of a total of 186 scintillating bars coupled with BCF-92 WLS fiber connected to Hamamatsu S13360 SiPMs. The vetoes detect positrons and electrons with momentum less than 450 MeV/c deflected by the magnetic field. Another veto detector is placed outside the magnetic field and close to the beam monitor system in order to detect positrons with momentum up to 500 MeV/c [4]. A TimePix3 silicon pixel detector placed just after the vacuum chamber exit window monitors the beam before the beam leaves the PADME setup. It provides information about position and time of every single particle in a bunch. The main PADME detector is the electromagnetic calorimeter (ECAL) [5]. It is placed at a distance of 3.46m with respect to the target. The ECAL detects the gammas in the final state. It has a ring-shape and is made of 616 BGO crystals. The scintilation-light decay time of $\sim$ 300ns of the BGO crystals determines the necessity of

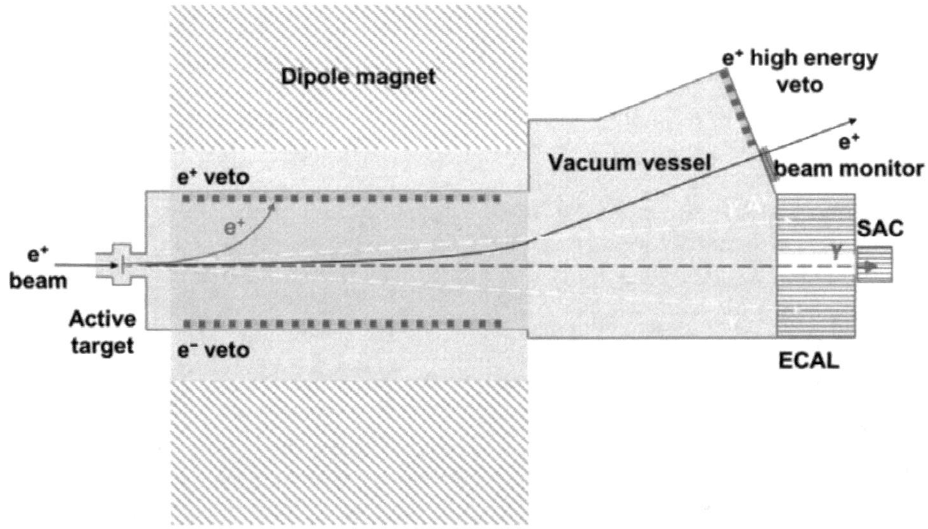

Figure 1. Schematic layout of the PADME experiment

a second detector with much faster response in order to be able to detect the higher rate of bremsstrahlung gammas, emitted at small angles with respect to the positron beam direction. Behind the hole at the center of the ECAL, a Small Angle Calorimeter (SAC) is positioned. It is made of 25 lead fluoride (PbF_2) crystals and provides angular coverage of [0, 20] mrad [6]. SAC is able to distinguish particles with time separation of $\sim$ 3 ns.

2.2. Background estimation

The first data-taking run was performed between October 2018 and February 2019. The main goal was to perform detector commissioning, background estimation and beam tuning. The main physics sources of background are the multiphoton annihilation process $e^+e^- \to \gamma\gamma(\gamma)$ and bremsstrahlung events $e^+N \to e^+N\gamma$. During the first run of PADME a higher level of the beam induced background was observed. It lead to increased total energy deposition in the ECAL. The first part of this run was performed with 545 MeV secondary beam, 25k e^+ on target and 250 ns bunch length. To reduce the background, the second part of the run was performed with 490 MeV primary beam with 25k e^+ on target and 250 ns bunch length. Additional Monte-Carlo studies showed that background events were also formed due to unwanted interaction between the beam and a Beryllium window, which at the time was separating the accelerator's vacuum from the detector's vacuum. This caused an extra scattering of the beam positrons, creating secondary particles that are not well centered with the beam axis and therefore showering the ECAL randomly. This problem was solved by installing a new mylar window more upstream the beam-line. During the second data acquisition run, an improved primary 430 MeV beam with 28k e^+ on target and 280 ns bunch length was used [7, 8]. The results of reducing the background index are shown on Table 1.

The improvement of the beam conditions and the reconstruction power of PADME are also visible in fig. 2 where the sum of the energy of two photons, detected in the ECAL, is shown. The $E_{\gamma\gamma} = E_{\gamma_1} + E_{\gamma_2}$ peak at the beam energy corresponds to $e^+e^- \to \gamma\gamma$ annihilation events. In the case of secondary beam the distribution is entirely dominated by the beam induced background while the substitution of the beam window with mylar leads to improvement in the resolution

Table 1. Beam induced background for different running conditions

Beam conditions	Background
RUN I: 545 MeV Secondary beam	0.36 MeV/e^+
RUN I: 490 MeV Primary beam	0.03 MeV/e^+
RUN II: 430 MeV Primary beam, Mylar window	0.013 MeV/e^+

and additional suppression of the events at lower $E_{\gamma\gamma}$.

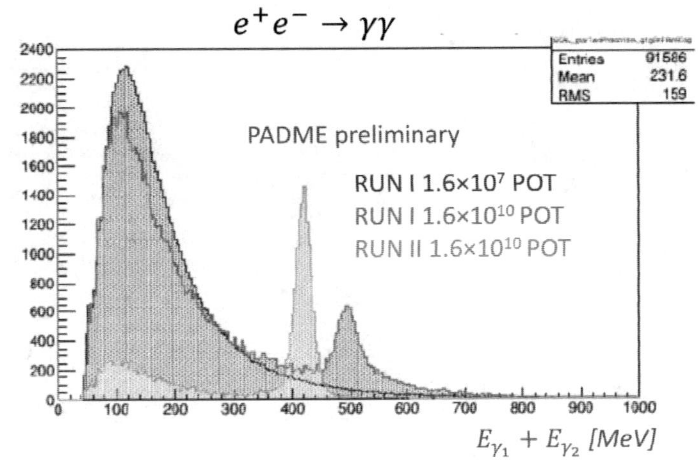

Figure 2. Energy distribution for two γ final state reconstructed events in the ECAL at different conditions

3. Additional dark sector studies

The dark photon is not the only dark sector particle in the reach of PADME parameter space. An attractive result was presented from the ATOMKI [9] group, pointing towards the existence of a new light boson with a mass $m_X \sim 17$ MeV/c^2, the so called X17. Such a particle is exactly in the PADME sensitivity region. In addition, a dedicated run with a $E = 282.779$ MeV positron beam on a target can lead to a resonant production of this hypothetical particle. However, due to the absence of a recoil photon in the resonant production, the current setup needs an improvement in order to be suitable for such study. X17 decays into leptons in the final state and precise information about their momentum and position is needed. Currently the PADME veto detectors are acting more or less as a hodoscopes, rather than spectrometers. Additional beam tuning and a proper target selection are also part of the setup improvement for the next stage of the experiment, where the main goal will be to search for X17.

Another new physics candidate within the reach of PADME are Axion-Like Particles (ALPs). ALPs are light pseudoscalar particles, which can also act as a portal between the dark sector and SM. Unlike the QCD axion, their mass and coupling constant are independent, which gives a wider region of parameter space. Both visible and invisible decay modes can be exploited by PADME. The ALPs production mechanism is similar to the dark photon's one, while the decay modes can be $ALP \to \gamma\gamma\gamma$, $ALP \to e^+e^-$ (for this channel the PADME sensitivity is at maximum), or the ALP itself can be a long-living particle. As in the case of the dark photon, the signature of an ALP is a peak in the missing mass distribution.With the current setup PADME aims to search for ALPs with masses in the interval 2 - 23.7 MeV/c^2 [10].

Dark higgs studies are also a part of the future PADME physics program. The model of interest for PADME is with $m_{h'} > 2m_{A'}$, where $m_{h'}$ is the Dark higgs' mass. The production mechanism is given by $e^+e^- \to h'A' \to 3(e^+e^-)$. The six leptons final state indicates the importance of the charged particle detectors to be used as spectrometers. The Dark higgs production cross-section for beam energy E = 500 MeV was computed to be $\sigma(h'A') \sim 1000$ pb for $\epsilon^2 \sim 10^{-6}$, which for an year of data taking with RUN II instantaneous luminosity can give the production of 150 h' mediated e^+e^- in six-leptons final states. For $e^+e^- \to 3(e^+e^-)$, we have $\sigma_{e^+e^-} \to 3(e^+e^-) \sim 1500$ pb, which indicates that for one year data taking with RUN II instantaneous luminosity can give the production of 200 SM $e^+e^- \to 3(e^+e^-)$ events. Even a modest background rejection, obtained with the present PADME veto system, will allow the dark sector to become the dominant process [11].

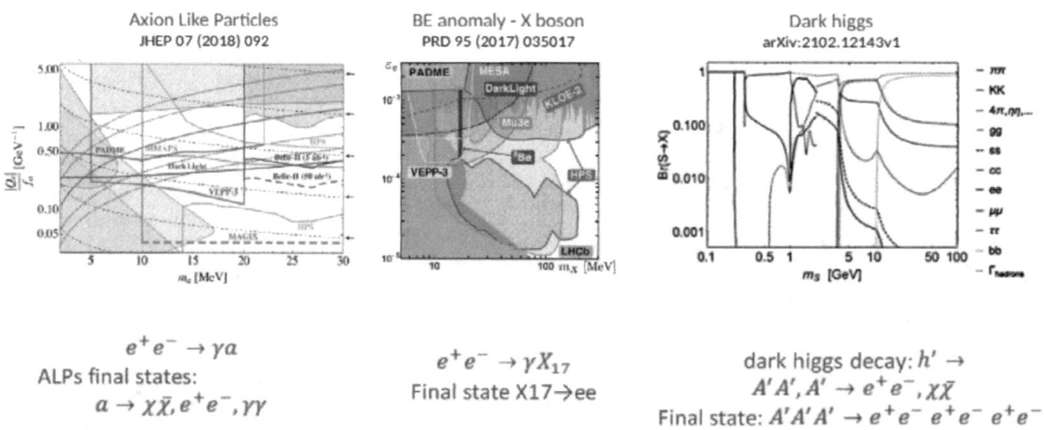

Figure 3. Possible regions for additional dark sector particle studies [12] [13] [14]

The PADME sensitivity to the presented scenarios is illustrated in fig. 3. Both ALP and the X17 particles are within the PADME reach, for specific unexplored model parameters.

4. Conclusions

The incompleteness of the Standard Model points towards the need of a new fundamental theory in order to understand the nature of the dark matter. A possible solution is a set of new particles of an additional dark sector. PADME aims to study the dark photon as a mediator between the dark sector and the Standard Model leptons. The experiment explores the process $e^+ + e^- \to \gamma + A'$ in which the presence of a dark photon should result into a peak in the missing mass distribution. Two data acquisition runs were performed. PADME RUN-I helped to define the background of the experiment, calibrate the detectors and improve the set-up. The analysis of PADME RUN-II is ongoing. Future improvements of the experiment may shed even more light on the dark sector particles.

Acknowledgments

The work is partly supported by MUCCA, CHIST-ERA-19-XAI-009 BG-NSF KP-06-DO02/4 from 15.12.2020 and MoU SU – LNF-INFN 70-06-497/07-10-2014.

5. References

[1] Mauro Raggi, Venelin Kozhuharov, *Proposal to Search for a Dark Photon in Positron on Target Collisions at DAΦNE Linac*, Adv.High Energy Phys. 2014 (2014), 959802.

[2] P. Valente, M. Belli, B. Bolli, B. Buonomo, S. Cantarella, R. Ceccarelli, A. Cecchinelli, O. Cerafogli, R. Clementi and C. Di Giulio, *et al.* [arXiv:1603.05651 [physics.acc-ph]].

[3] F. Oliva, Operation and performance of the active target of PADME, Nuclear Inst. and Methods in Physics Research, A (2019) 162354,https://doi.org/10.1016/j.nima.2019.162354

[4] G. Georgiev, S. Ivanov, V. Kozhuharov, M. Mitev, R. Simeonov and L. Tsankov, "Performance of the Front-End Electronics of the PADME charged particle detector system," 2018 IEEE XXVII International Scientific Conference Electronics - ET, 2018, pp. 1-4, doi: 10.1109/ET.2018.8549581.

[5] P. Albicocco, *et al.* JINST **15** (2020) no.10, T10003 doi:10.1088/1748-0221/15/10/T10003 [arXiv:2007.14240 [physics.ins-det]].

[6] A. Frankenthal, *et al.* Nucl. Instrum. Meth. A **919** (2019), 89-97 doi:10.1016/j.nima.2018.12.035 [arXiv:1809.10840 [physics.ins-det]].

[7] P. Valente et al., J. Phys. Conf. Ser. 874, 012017 (2017)

[8] A. Frankenthal, "Searching for Dark Photons with PADME," [arXiv:1910.00764 [physics.ins-det]].

[9] A. J. Krasznahorkay, M. Csatlós, L. Csige, Z. Gácsi, J. Gulyás, M. Hunyadi, T. J. Ketel, A. Krasznahorkay, I. Kuti and B. M. Nyakó, *et al.* Phys. Rev. Lett. **116** (2016) no.4, 042501 doi:10.1103/PhysRevLett.116.042501 [arXiv:1504.01527 [nucl-ex]].

[10] L. Darmé, F. Giacchino, E. Nardi and M. Raggi, JHEP **06** (2021), 009 doi:10.1007/JHEP06(2021)009 [arXiv:2012.07894 [hep-ph]].

[11] P. Ciafaloni, G. Martelli and M. Raggi, "Searching for dark sectors in multi lepton final state in $e^+ e^-$ collisions," JHEP **04** (2021), 163 doi:10.1007/JHEP04(2021)163 [arXiv:2012.04754 [hep-ph]].

[12] Alves, D.S.M., Weiner, N. A viable QCD axion in the MeV mass range. J. High Energ. Phys. 2018, 92 (2018). https://doi.org/10.1007/JHEP07(2018)092v

[13] J. L. Feng, B. Fornal, I. Galon, S. Gardner, J. Smolinsky, T. M. P. Tait and P. Tanedo, Phys. Rev. D **95** (2017) no.3, 035017 doi:10.1103/PhysRevD.95.035017 [arXiv:1608.03591 [hep-ph]].

[14] P. Agrawal, M. Bauer, J. Beacham, A. Berlin, A. Boyarsky, S. Cebrian, X. Cid-Vidal, D. d'Enterria, A. De Roeck and M. Drewes, *et al.* [arXiv:2102.12143 [hep-ph]].

Feasibility Study of Simultaneous Measurement of Multiple Parameters from a Single Monolithic Scintillator Crystal Using the Geant4 Simulation Toolkit

S Ivanov and Zh Toneva

Faculty of Physics, Sofia University, 5 Bourchier Str. 1142 Sofia, Bulgaria

E-mail: simo@phys.uni-sofia.bg

Abstract. A feasibility study for the construction of a minimal multi-parameter detector module from a single monolithic $CeBr_3$ scintillator crystal and a SiPM matrix that is capable of simultaneously measuring deposited energy, position of particle interaction and time characteristics was performed by simulation and direct measurements.

The results show that the detector module can operate successfully with characteristics that are close to the limits of the physical detector and that such minimal modules can potentially be used on micro- and nano-satellite platforms .

1. Introduction

The ability to observe transient events and perform gamma-ray imaging and spectroscopy in the sub-MeV and low MeV ranges are of considerable scientific interest and importance for astrophysics, for example in the research of supernovae explosion dynamics, nucleosynthesis radioactivity, annihilation radiation, etc. [1, 2]. The availability of low-cost space-based observatories with performance on par with terrestrial detectors with similar specifications is key to achieving this ability.

The primary goal of this research is to investigate applications of recent developments in detector, sensor and satellite technologies in the field of gamma imaging and spectroscopy and the desired energy ranges. In addition to possessing the necessary detector characteristics, such instruments must fit within the technological limitations of the satellite platforms targeted as deployment vehicle.

New generation scintillators and semiconductor photon counting devices were identified as a promising direction previously [1].

This article describes an approach and preliminary simulation results for the performance of a combined position and energy sensitive detector built from a single crystal scintillator and a SiPM readout matrix.

2. Module Models

The strict platform limitations of most micro and nano-satellite platforms (e.g. [3]) necessitate development of minimalistic detector configurations with the lowest possible size, weight, power and thermal output characteristics and sufficient mechanical stability.

Consequently, commercial, off-the-shelf cylindrical and rectangular prism scintillators from several suppliers were chosen for a feasibility study of the operation of a multi-parameter detector. A review of the literature appears to indicate little work and few results using this approach.

Cerium Bromide was chosen as primary scintillator material because of its high photon yield and very good intrinsic energy resolution as well as its excellent timing characteristics [4].

Figure 1. Cylindrical model with reflector.

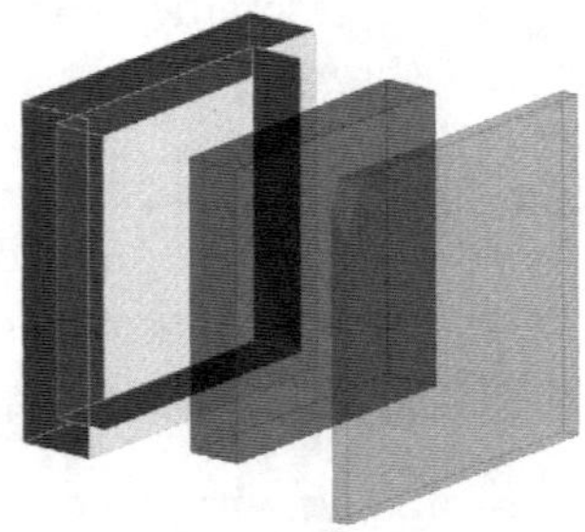

Figure 2. Prismatic model with reflector.

Two basic parametrized modules were developed for a cylinder and a prism scintillator (see figure 1 and figure 2), with and without reflective cases, and a number of simulations were performed to evaluate the possibility for readout of multiple parameters from them for different gamma ray sources.

The module parameters are the size and the shape of the scintillator crystal and its material properties - chemical composition, scintillator and optical characteristics, dimensions of its enclosure and optical window and their surfaces' optical properties. The configuration of the module also includes parametrized definitions of the attached sensitive detectors, the quantities they track and the output configuration.

3. Approach to simulation

The Geant4 toolkit [5] is widely used to build detector geometries and simulate detector operation. It provides the end user with several methods for geometry construction, among them construction in C++ code, and import of CAD designs, realized by various third party libraries, and construction from an XML description using a markup subset, defined by the GDML specification.

The requirements of the project necessitated a large number of simulation tests of many modules of similar shapes with minor differences in their geometries, enclosure construction details and materials. A natural approach would be to create several parametrized and configurable modules, and test these by changing their configurations at simulation run-time.

We found that all geometry construction methods available to the simulation developer by default present shortcomings that would require significant implementation effort.

Of the three available approaches, the development of models via the geometry description markup language was seen as the best option, although it still has several shortcomings that make the repetitive construction of modules as described above difficult.

The most significant issues are the lack of locally scoped variables, and the lack of ability to treat complex logical structures as units. The suggested approach by the GDML documentation is to re-implement the parser and detector construction module, which is a non-trivial task, and

Journal of Physics: Conference Series **2255** (2022) 012010 doi:10.1088/1742-6596/2255/1/012010

requires significant investment for development and maintenance time that is difficult to justify for small projects with limited scope.

To overcome these issues for this project, the necessary additional functionality was added as a custom extension of the GDML language, and an xquery preprocessor was developed which reads this extended geometry, applies loops and transformations, including private variable substitutions and xml component definitions into a standard GDML file, which can subsequently be read by the unmodified G4 parser and executed.

This approach allows the simulation system to build complex detectors from simple, "black-box" flexible and configurable component definitions without being limited by the G4 implementation. The approach was used to study several detector geometries for applications in line with the project's goals.

In addition, our implementation allows materials, surface properties and other features of the GDML specification to be defined in libraries and reused in detector geometry construction as parameters to the respective parametrized components.

Finally, a simple GDML extension allows the system to load, initialize and attach sensitive detectors to parts of the detector geometry through the GDML geometry descriptor, removing the need for geometry building and sensitive detector construction and initialization code completely.

The simulation package has also implemented a streamlined, XML-configurable data extraction system, which has reduced significantly the interdependencies between user action modules, detector geometry and sensitive detector code. A data output API has been implemented, which allows all data extraction to be abstracted into one G4 RunUserAction subclass and the specific SensitiveDetector subclasses and to decouple data extraction completely from the rest of the simulation.

4. Preliminary Results

Several experimental studies were performed on typical $CeBr_3$ scintillators and matched with simulations on their counterpart modules to confirm the concept and validate the operation of the simulation system. The studied scintillators are $50 \times 50 \times 10mm^3$, $50 \times 50 \times 25mm^3$ rectangular prisms and $25.4 \times 25.4mm^3$ cylinder with reflector casings and optical window on the base.

Energy deposition, optical photon distribution and time of arrival statistics were collected with a SiPM matrix model attached to the optical window of the crystal.

Table 1. Comparison of experimentally obtained and simulated resolution measurement for a $CeBr_3$ prismatic scintillator.

$Energy\,[keV]$	$Experimental\,Resolution\,[\%]$	$Simulated\,Resolution[\%]$
122	5.0	6.4
244	4.2	5.6
344	3.7	4.7
778	2.4	3.0
1408	3.0	4.0

Energy resolutions for several spectral lines from measured and simulated spectra are summarized in table 1. The results are in good agreement with the specifications of the crystal and with similar measurements performed elsewhere.

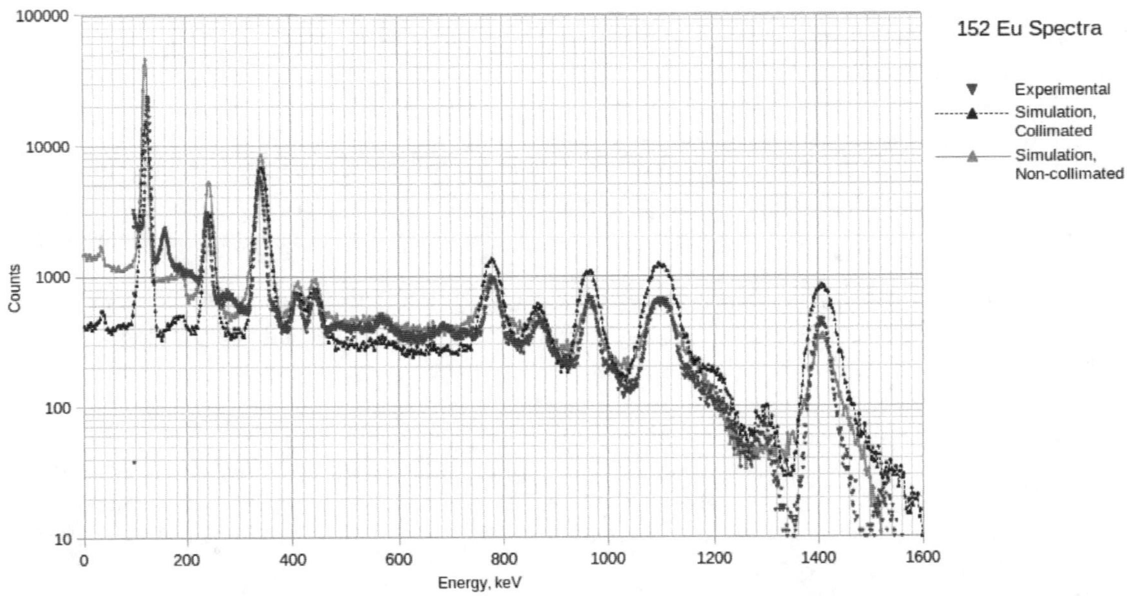

Figure 3. Comparison of simulated and experimental ^{152}Eu spectra for $CeBr_3$ scintillator.

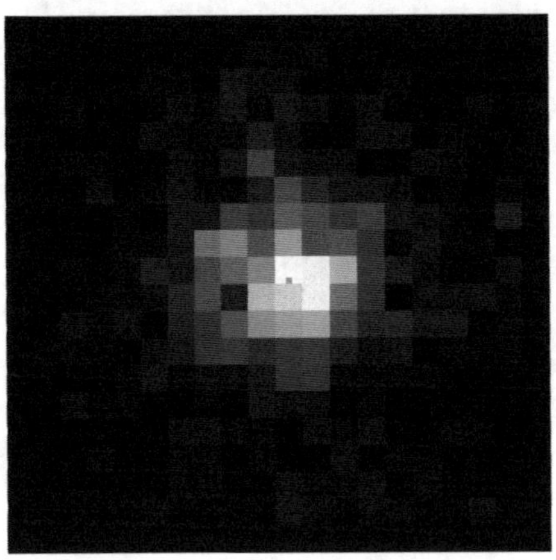

Figure 4. Optical photons distribution, actual and calculated interaction position for a hit near the detector center.

Figure 5. Optical photons distribution, actual and calculated interaction position for a hit near the edge of the detector.

Figures 4 and 5 show typical simulated response, registered by the SiPM matrix. Each square represents one matrix element (one silicon photomultiplier receptor); the colored dots represent the actual interactions and the estimated positions, derived from the intensity distribution registered by the individual SiPM units. The orange dots represent the interaction (as registered by the respective SensitiveDetector class), and the green dots represent the estimated interaction positions. The different levels of gray represent the scaled intensity of incident photons, registered by the individual SiPM cells, where the actual measurements are distributed into 256 levels of gray. Individual cells in a single picture are comparable, but intensities in different pictures are not.

Simultaneously with the energy measurements, the interaction positions were estimated from the distribution of photon counts on the individual pixels of several SiPM matrices of different sizes and layouts using a variant of the centroid method [6].

Based on several simulation runs of 10k events each, the position of the interaction is estimated with x, y-position resolution of 2.05 mm and σ of 1.1 mm for a $50 \times 50 \times 25\ mm^3$ matrix built from $3 \times 3\ mm^2$ SiPMs without accounting for edge reflections. Using a naive approach of removing the pixels closest to the edges in the calculation of the position improved the position resolution to 1.56 mm with a standard deviation σ of 0.88 mm with a decrease in detection efficiency.

The reason for this improvement is the redistribution of intensity by reflections from the walls of the scintillator, which skew the photon distribution, but are essential for the determination of the deposited energy.

5. Conclusion

A feasibility study for the construction of a minimal multi-parameter detector module from a single monolithic scintillator crystal and a SiPM matrix that is capable of simultaneously measuring deposited energy, position of particle interaction and time resolution was performed by simulation and direct measurement.

The results show that the detector module can operate successfully with characteristics that are close to the limits of the physical detector. Using this approach, concurrent measurements of position of interaction and deposited energy were performed with good accuracy. The research validates the proposed concept of minimalistic detector modules that can potentially be used on the targeted satellite platforms of interest or other resource-limited scenarios.

Acknowledgements

The author acknowledges partial support by BG-NSF under grant MON-FNI DN18/17.

References

[1] Toneva Z 2019 *Springer Proceedings in Physics* **15** 247
[2] Kurfess J D 2000 *AIP* **510** 789
[3] Cubesat 2014 Cubesat design specification rev.13 online
[4] Toneva Z 2020 *J. Inst.* **15** C01013–C01013
[5] Agostinelli S 2003 *Nuclear Instruments and Methods in Physics Research Section A: Accelerators, Spectrometers, Detectors and Associated Equipment* **506(3)** 250–303
[6] Ivanov S 2021 *Simulation of Single-crystal Compton Imaging Camera* Master's thesis Sofia University

Impact Point Determination of Gamma Interaction in a Monolithic Scintillator Crystal Read Out by a SiPM Matrix

Svetoslav Ivanov and Simeon Ivanov

Faculty of Physics, Sofia University, 5 James Bourchier Blvd., 1142 Sofia, Bulgaria

E-mail: `svvetlio@phys.uni-sofia.bg`

Abstract.
We construct a simulation framework which can determine simultaneously deposited energy, position and time resolution of interactions of gamma rays in a monolithic scintillator crystal read out by a matrix of photosensitive detectors. And we report and discuss the simulation results form three different methods to determine the impact position of a particle.

1. Introduction

Research of phenomena in several areas of modern astrophysics like supernovae explosion mechanisms, nucleosynthesis radioactivity, annihilation radiation and others depends on the capability to observe transient events with high time resolution and perform gamma ray imaging and spectroscopy in the sub-MeV and low MeV ranges.

The goal of this research is to investigate modern detector technologies and design detector modules that would allow development of light and compact instruments, suitable for deployment on low-cost satellite platforms like CubeSat [1], within the strict limitations inherent in those platforms.

The approach that is being researched is the feasibility of construction of modules with multiple detection capabilities (energy, position and time of interaction) and high resolution [2]. In this article we examine several methods for determination the position of a gamma interaction.

2. Approach

To study the interactions for different geometries and readout patterns, a simulation application was designed and built with the G4 framework [3]. The application allows building of complex detector geometries from parametrized predefined XML snippets, which can be combined and positioned using a preprocessor and extended GDML-like syntax. Additionally, a configurable system for data collection and readout has been added to remove typical interdependencies and make development of extension libraries simpler and less resource consuming. An application to simulate the analog electronic processing and a set of analysis filters is in development.

Several prospective technologies for the development of the targeted detector modules have been reviewed and compared with regard to their capabilities and operating requirements. Several recently developed scintillators with high photon yields, high energy resolution and good timing characteristics have been identified as especially promising candidates for primary detectors [4]. Silicon photomultipliers have been selected as the technology for readout electronics, and a hybrid approach for signal processing based on custom fast analog amplification, amplitude to time conversion and FPGA-based analog to digital conversion is being developed. The feasibility of embedding of some analysis and reconstruction algorithms in the detector system is also researched.

A set of parametrized simulation modules of all technology candidates has been developed and a simulation application pipeline has been designed and partially implemented for the full detector processing chain, from the detector physical process to electric signal processing and analysis software. A hardware validation platform is simultaneously being designed to validate and implement a prototype.

This article discusses some implementation details and preliminary results of the data processing and analysis components of the proposed detector system.

3. Impact point determination

We examined three methods to determine the position of the initial gamma interaction in a monolithic scintillator crystal read out by a SiPM matrix.

3.1. Center of Mass

The first method is the center of mass [5]. It determines the impact point in the plane defined by the SiPM matrix by the weighted average light intensity of the distributed photons produced by the incident particle. This point is calculated by the following equation

$$x_j = s\frac{\sum_{i=1}^{n} iI_i}{\sum_{i=1}^{n} I_i} \quad (1)$$

where x_j is the position of interaction, s is the size of a pixel in the matrix, I_i is the light intensity registered in a pixel and n is the number of pixels on every dimension of the detector.

We simulated a number of different sizes, layouts and materials of the scintillator crystal and SiPM matrix and different energies of the gamma particles. For a $50 \times 50 \times 25mm^3$ $CeBr_3$ scintillator read out by a $50 \times 50mm^2$ matrix built from $3 \times 3mm^2$ SiPM pixels, the position of interaction is estimated with resolution of $2.05mm$ and a standard deviation of $1.1mm$ [6] compared to the simulated position of interaction.

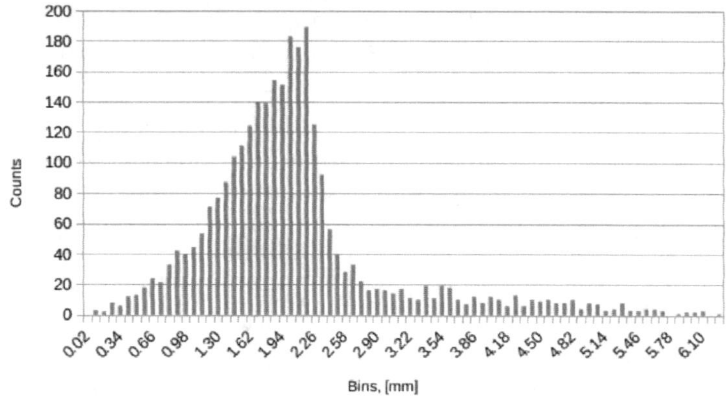

Figure 1. Distribution of the difference between the position of the interaction and the estimated position.

Such a scintillator crystal and SiPM matrix are chosen because of their market availability and possible experimental validation of the simulation results.

3.2. Detection of Multiple Interactions

In the second method, we estimate the position of incident gamma while take into account that there may be more than one interaction in the crystal volume. We construct a method that searches for a multiple maxima in the light distribution over the plane defined by the SiPM matrix. The algorithm finds the local maxima in every column and row and then distinguishes the real maximum from a local small random increase in the light intensity. Once we find such a pattern, we reconstruct the position of the gamma particle by weighting the parameters of these multiple points and neglect the rest with a predefined threshold.

With the same simulated data set as the center of mass, this method shows increase in the accuracy for a multiple interaction events and for events positioned near the edges of the crystal. However, the overall estimation of the resolution of the position of impact is $2.37mm$ with standard deviation of $1.3mm$.

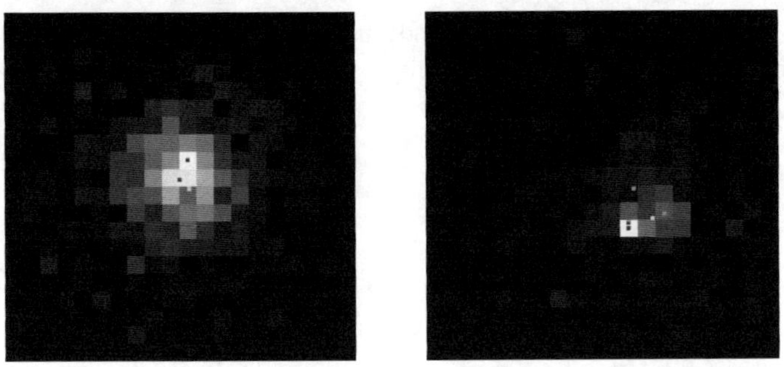

Figure 2. Optical photons distribution, actual (blue) and calculated interaction position via center of mass (green) and multiple interaction (violet) methods for a hit with (right picture) and without (left picture) multiple secondary interactions (Compton scattering in orange).

3.3. Analysis with ML methods

We examine a simple machine learning methods to estimate the position of the gamma particle. We approach two types of ML algorithms. In the first one, we build models based on the image classification methods which classify the event and then decide the optimal method to calculate the impact position with.

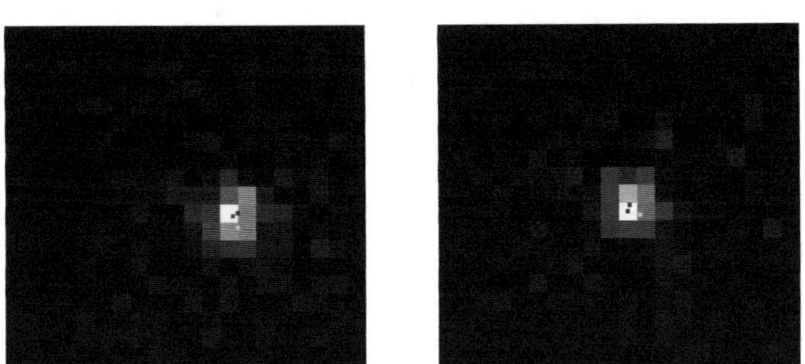

Figure 3. Optical photons distribution, actual (blue) and calculated interaction position via center of mass (green) and ML (violet) methods for a hit without multiple secondary interactions.

Journal of Physics: Conference Series **2255** (2022) 012011 doi:10.1088/1742-6596/2255/1/012011

In the second one, we create a simple image recognition algorithm based on a sample data including both single and multiple gamma interaction events. It searches for a pattern in the light distribution based on the training data set from the simulation which provides information for the type and energy of the particle, and the types and numbers of interactions.

4. Conclusion and Discussion

A framework which simulates the electronics output of a SiPM matrix and three methods for determination of position of a interacting particle are developed and evaluated. The evaluation shows that interaction position can be determined with accuracy of $2.05mm$ via a center of mass calculation.

Compared with the center of mass method, the multiple maximum method shows improvement in accuracy when the position of interaction is close to the edges of the scintillator crystal and reflections from the edges take significant role in the light distributed over the photosensitive matrix.

At this stage of development, the results for the position of the interacting gamma particle estimated by the ML algorithms are in a good agreement with the other two methods. We consider these methods to be useful in the space environment and when there is no good educated guess for the type and the energy of the interacting particle in the detector crystal.

Acknowledgements

The author acknowledges partial support by BG-NSF under grant MON-FNI DN18/17.

References

[1] Cubesat 2014 Cubesat design specification rev.13 online
[2] Ivanov S 2021 *Bulgarian Journal of Physics* **47** 31–41
[3] Agostinelli S 2003 *Nuclear Instruments and Methods in Physics Research Section A: Accelerators, Spectrometers, Detectors and Associated Equipment* **506(3)** 250–303
[4] Toneva Z 2019 *Springer Proceedings in Physics* **15** 247
[5] Bushinin 1973 *NIM* **106** 493–498
[6] Ivanov S 2021 *Simulation of Single-crystal Compton Imaging Camera* Master's thesis Sofia University

Extra-terrestrial influence on climate variability

Ts P Velichkova and N A Kilifarska

Climate, Atmosphere and Water Research Institute, Bulgarian Academy of Sciences, 66 Tsarigradsko Shosse Blvd, 1784 Sofia, Bulgaria

tsvetelinavelichkova@abv.bg

Abstract. This study is focused on the analysis of signatures of extra-terrestrial influence on the centennial variability of some climatic variables, during the period 1900-2010. The applied lagged cross-correlation analysis shows evidence for a synchronisation between intensity of galactic cosmic rays (GCR) and two climate variables – the air surface temperature and the sea level pressure. The time delay of climatic response is not more than 5 years in the regions of stronger correlations. Moreover, the centennial variability of GCR is imprinted on the ozone mixing ratio at 70 hPa, suggesting its role of a mediator of GCR influence on climate. Having in mind the ozone's capability to control the near tropopause temperature and humidity, we have suggested a mechanism describing the transmission of GCR signal from lower stratosphere, down to the Earth's surface.

1. Introduction

One of the most debatable topic in climatic research is related to the contemporary and historical changes of Earth's climate. The notion of the decisive role of anthropogenic gases is widespread among the scientific community [1-3]. However, according to another point of view, climate variability may be induced by factors outside the climatic system like variations in the solar activity, the changes in Earth's orbital parameters, the influence of the charged particles from a cosmic origin [4-6], etc. This section presents an overview of the potential mechanisms of influence of external factors on climate, being considered as important for the life on the Earth.

The changes in the main characteristics of the Sun are determined by its growth as a star, and by variations in solar magnetic field [7]. The effects of solar variability on climate are discussed long ago [8-10], due to the fact that signals of solar activity are detected in variations of different climatic variables [11-13]. However, the amplitude of variations in solar luminosity is very small – at 11-year, as well as on longer time scales [14, 15]. This suggests that in the atmosphere should exist some amplifying mechanism, which is able to magnify and mediate downward to the surface the initial solar "signal". For this reason, it is important to find out the physical processes involved in solar-climate coupling for the proper understanding of climate variability.

Some authors suggest that the 11-year periodicity of sunspot numbers, corresponding to periodical activation of the zonal (poloidal), or meridional (toroidal) component of the Sun's magnetic field could be related to the phase transition of regional climatic modes like North Atlantic oscillation (NAO), El Nino Southern Oscillation (ENSO), etc. [16, 17].

Besides the 11-year periodicity of solar magnetic field, there is another 22-year periodicity, related to the change in orientation of the solar magnetic dipole – from N→S to S→N direction. Sign of this periodicity in Earth's climate parameters have been found in the surface temperature [18, 19], precipitation [20], lower stratospheric ozone [21, 22], Earth rotation [23, 24], etc. However, the physical mechanism behind these relationships is not well understood [25].

The interference between gravitational force of Sun and the giant planets could infer some changes in orbital parameters of our planet. The first who pay attention on such a possibility has been the Serbian scientist Milutin Milankovitch, who suggested that: (i) the changes in the shape of Earth's revolution around the Sun (called *eccentricity*), together with (ii) the tilt of the axis of planetary rotation regarding the orbital plane (called *obliquity*) and (iii) its slow wobble (known as *precession*) could be an explanation of the periodic warming and cooling of Earth's climate [26].

Changes in orbital parameters, related to the eccentricity, ranges between 95 000 – 125 000 years, and modify the planetary orbit from circular to elliptical. The unequal distance from the Sun –in the case of elliptical orbit – altering the amount of radiation received, is given as an argument for observed climatic changes. The periodicity related to the precession of Earth's rotational axis is lasting ~ 26 000 years. It alters the amount of received radiation between summer and winter – especially important for the polar regions. The last cycle, associated with the tilt of the planetary axis of rotation, has a period of ~ 41,000 years. The tilt is changing between 22° and 24°. The higher angle increases the amplitude of seasonally varying insulation received from the Sun, in both hemispheres.

Another extra-terrestrial factor, with a possible influence on climate, is the flux of galactic cosmic rays (GCR), continuously penetrating in Earth's atmosphere. The authors of [27] have reported that the intensity of GCRs during solar minimums 23/24 and 24/25 is the highest ever recorded. The flux of GCR reaching the Earth is modulated by the solar magnetic field in Heliosphere, as well as by the strength and heterogeneity of geomagnetic field. There are several suggested mechanisms, aimed to explain the GCR influence on climate, and practically all of them are focused on the compositional changes induced by GCR in the atmosphere. According to [28] the intensity of GCR could increase amount of NO_2 in the middle atmosphere and in such a way to reduce the solar radiation reaching the planetary surface. The effect is expected to be stronger at the polar regions, where the concentration of precipitating particles is higher. The surface temperature and pressure should respond to the meridional gradient of differential heating [29], followed by changes of atmospheric circulation [30]. One of the most popular, however, is the possibility for GCR influence on the cloud cover, and respectively on the amount of the solar radiation received by the planet [31-33].

This paper presents some more evidence for an existing covariance between variability of GCR intensity and climatic variables, including air surface temperature, sea level atmospheric pressure and lower stratospheric ozone. A physical mechanism, explaining the synchronisation of the variances of examined variables, is also presented.

2. Data and methods used

The monthly mean data for air temperature at 2 meters above the surface (T_{2m}), sea level pressure (P), ozone mixing ratio at 70 hPa (O_3) and water vapour at 150 hPa (H_2O) are provided from ERA 20C reanalysis, covering the period 1900-2010. Winter seasonal means have been calculated from December - March monthly values. The data have been smoothed by an 11-point running average procedure.

We have used annual values of galactic cosmic rays (GCR), provided by the *World Data Centre for Paleoclimatology*, Bolder, and the *NOAA Paleoclimatology Program*. The record has been expanded since 2007 to nowadays by the Moscow neutron monitor data.

The temporal covariance between galactic cosmic rays, lower stratospheric ozone, near surface temperature (T_{2m}), and the sea-level pressure is estimated by the use of the lagged cross-correlation analysis. Due to the fact that the cross-correlation function is not symmetrical about the zero lag, in STATISTICA package the independent (i.e. first, or leading) variable is moved first backward, and then forward. The cross-correlation coefficient with lag k is computed following the standard formulas, as described in most time series references (e.g., [34]):

$$r_{xy}(k) = c_{xy}(k)/(s_x s_y), \quad \text{for k=0, ±1, ±2, ±3, ………} \qquad (1)$$

where c_{xy} is the cross-covariance coefficients at lag k. s_x and s_y are standard deviations of series X and Y respectively, and are calculated by the formulas:

$$s_x = \sqrt{\frac{\sum(X_t-\bar{X})^2}{N}}; \qquad s_y = \sqrt{\frac{\sum(Y_t-\bar{Y})^2}{N}} \tag{2}$$

where N is the number of the observations in the time series. The cross covariance function is calculated as follow:

$$c_{xy}(k) = \frac{1}{N}\cdot\sum[(X_t-\bar{X})\cdot(Y_{t+m}-\bar{Y})] \qquad \text{for t=1, N-1; m=0, 1, 2,... k} \tag{3}$$

$$c_{xy}(k) = \frac{1}{N}\cdot\sum[(Y_t-\bar{Y})\cdot(X_{t+m}-\bar{X})] \qquad \text{for t=1, N+1; m=-1, -2,... -k} \tag{4}$$

Equation (3) describes the delayed response of Y to changes of X in moment ***t***, and its time lag is given by a positive number. The Eq. (4) describes lagged response of X to the fluctuation of Y in the moment ***t***. For this reason, the time delay in the maps of lagged response could be a positive or negative number.

Correlation maps, illustrating the spatial irregularities in the strength of the covariance between analysed variables, have been created from the statistically significant at 2σ level correlation coefficients (calculated in each point of our grid, with an increment of 10° in latitude and longitude). Due to the relatively lower power of this statistical method some authors, e.g. [34], have recommended a replication of the calculated correlations at different time lags and different groups of subjects. For this reason, each lagged correlation coefficient has been selected to be the maximal value, among all statistically significant coefficients, calculated with time lags from 0 to 35 years. The correlation coefficients are statistically significant at α=0.05, meaning that probability of making error of type 1, when rejecting the null hypothesis (i.e. supposing that two variables are uncorrelated), does not exceed 5%.

However, the lagged correlation coefficients have different time lags in different points of our grid, which makes it difficult to compare them directly. For this reason, they have been preliminary weighted by the autocorrelation function of the forcing factor – with a lag corresponding to the response variable's delay to the applied impact. Although this procedure reduces the strength of the relation, it allows us to compare the impact of correlation coefficients with different time lags (for more details see chapter 12 in [34]). The reasoning for such a weighting is based on the assumption that the effect of the applied (in a given moment) forcing, decreases with moving away from this moment.

3. Results

In this section we have examined how and to what degree the long-term variations of intensity of GCR can influence the centennial variability of the air surface temperature at 2 m above the Earth's surface (T_{2m}) and sea level pressure (P).

We have performed a quantitative estimation of the relation between GCR and T_{2m} for the period 1900 – 2010 by the use of a lagged cross-correlation analysis. The correlation coefficients – describing the "GCR influence on the surface temperature" – have been weighted through multiplication of the value of the autocorrelation function of GCR corresponding to a given time lag [34]. We do this because the memory of the climate system for the applied impact weakens with time, which means that the high correlation coefficients with a large delay are more or less random. From the weighted correlation coefficients have been created correlation map, showing the spatial distribution of the degree of similarity between temporal variability of examined variables. The results are presented in figure 1.

Examination of figure 1 reveals that the strength of connectivity between GCR and T_{2m} is unevenly distributed over the globe. The correlation between both variables is mostly negative, being definitely stronger in Sothern Hemisphere middle latitudes. This result is in accordance with the observed geomagnetic field decrease in the Southern Hemisphere during the 20-th century [35, 36], allowing more particles to precipitate in the regionally weaker geomagnetic field. The delay of the T_{2m} response to the GCRs impact is less than 6 years in the regions of stronger correlation (see the bottom panel of figure 1).

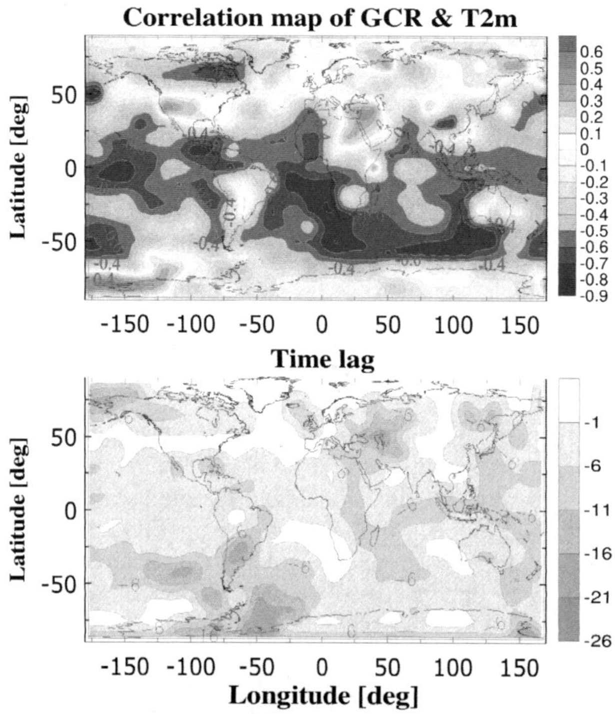

Figure 1. Correlation map of the spatial distribution of the degree of connectivity between temporal variability of GCR and T_{2m} during the period 1900 - 2010. Bottom panel present the time delay of the near surface temperature to GCR forcing in years.

An illustration of the uneven distribution of the connectivity between GCR and sea level pressure (P) for the e period 1900 – 2010 is shown in figure 2. The covariations of both variables is stronger in the Southern than in the Northern Hemisphere. In contrast to the equatorial Atlantic Ocean and the Far East, where the both variables covariate in-phase, at extra-tropics their variations are in antiphase. The bottom panel in figure 2 shows the time lag of the sea level pressure response to the GCR's forcing. Note that the pressure responds faster in regions with antiphase variations, compared to regions with in-phase covariation.

This coherence in the variations of climatic variables and GCR is an indication about a possible causality between these variable. The question is what is the mechanisms of translation of GCR influence down to the Earth's surface. A brief description of a possible one is given in the following section.

4. Mechanism for GCR influence on climate

According to [37, 38], the GCR flux is modulated by the solar wind and the geomagnetic field, whose spatial inhomogeneity focuses or defocuses the charged energetic particles in certain regions of the world. The ionization they create near the tropopause activates ion-molecular reactions that create or destroy ozone near the tropopause [39-41]. The strong ability of ozone to absorb electromagnetic radiation with different wavelengths, determines its capability to affect the stratospheric temperature. Of special importance is its influence on the neat tropopause temperature. The latter controls not only the radiative efficiency of atmospheric composition in this region, but also the static stability of the upper troposphere [42].

For example, ozone depletion in the lower stratosphere is accompanied by a cooling of the near tropopause region and destabilization of the upper troposphere [43]. The instability allows uplifting of the wetter air masses from the lower atmospheric levels and moistening of the upper troposphere.

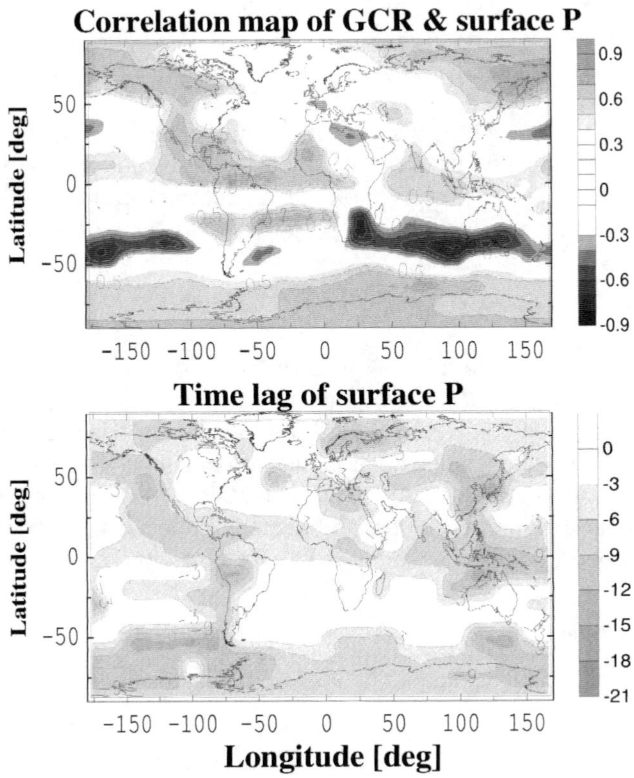

Figure 2. Spatial distribution of the correlation between temporary variable GCR and sea level pressure (P) during the period 1900 - 2010. Bottom panel present the time delay in years of the sea level pressure to GCR forcing.

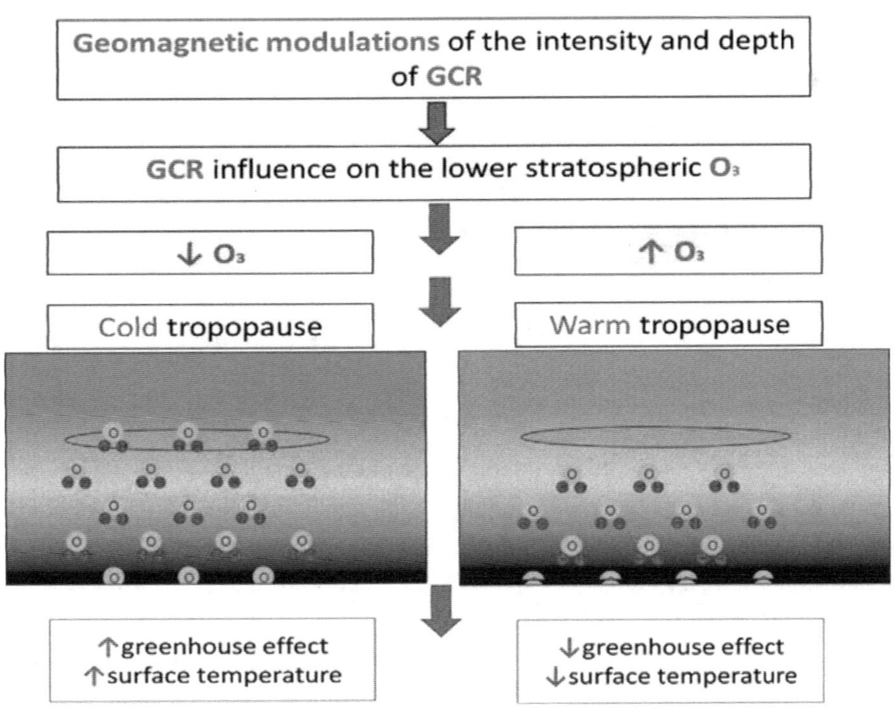

Figure 2. Scheme of the mechanism of galactic cosmic rays (GCR) influence on the Earth's surface.

In addition, the cold water vapour has extremely low emissivity, thus keeping the absorbed long-wave radiation in the troposphere and increasing the greenhouse effect. In this context, satellite measurements show that 90% of the greenhouse power of the atmospheric water vapour is due to the near tropopause water vapour [44].

On the other side, ozone enhancement in the lower stratosphere warms the near tropopause region, thus increasing the static stability of the upper atmosphere and putting it in a mode of gradual dehydration [45]. Saying it differently, O_3 deficit near the tropopause is projected on the surface as warming, while ozone abundance is followed by cooling of the surface temperature. The so described chain of links is schematically presented in figure 3.

5. Validation of the mechanism proposed

A kind of validation of the above hypothesis is shown in figure 4. The first link of the causal chain of relations, describing GCR influence on climate, consists of GCR impact on the lower stratospheric ozone density. So we have examined the strength of connectivity between GCR and ozone and its distribution over the planet. The left panel of the figure 4 presents the map of centennial covariance between GCR flux reaching the ground and winter ozone at 70 hPa. Note that at tropical and polar latitudes, the GCR and O_3 are covarying in-phase. This means that the centennial decrease of GCR intensity, during the 20-th century, is accompanied by a diminishing of the lower stratospheric ozone in these regions.

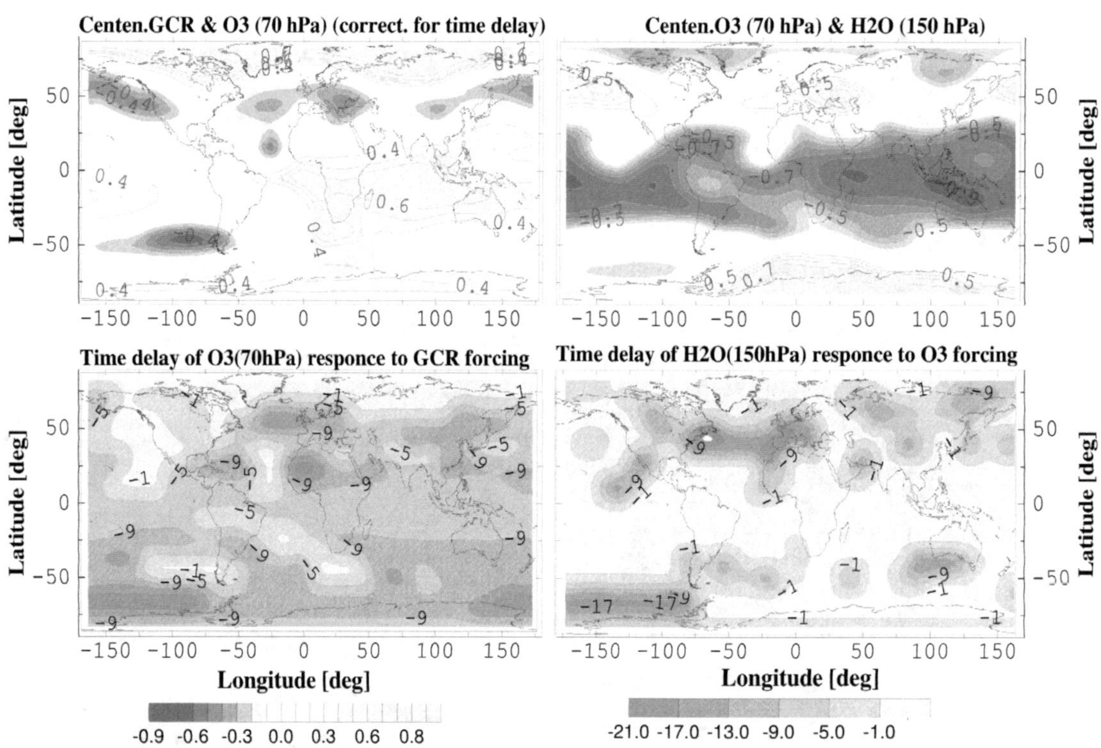

Figure 3. Correlation map of (left) the galactic cosmic rays (GCR) and winter ozone at 70hPa and (right) ozone's influence on the water vapour at 150 hPa. Bottom panels show the time delay of O_3 response to GCR forcing (left) and water vapour reaction to O_3 forcing (right).

On other hand, the correlation between both variables is negative in some regions near 55°N and 55°S latitudes, suggesting an ozone enhancement there. At first glance the both results seem controversial, but it has to be stressed that the two regions are influenced by particles with different energies. For example, the polar regions are affected mainly by GCR – freely penetrating along the open geomagnetic field lines. These particles are modulated mainly by the variations of helio-magnetic field. The latitudes near the 50^0, however, are mostly impacted by the trapped particles in the Earth's radiation belts, which are controlled entirely by the geomagnetic field. In addition, the different energies and depth of penetration in the atmosphere, are the main reason for activation of different ion-molecular reactions, creating or destructing the atmospheric ozone [38, 45].

The analysis of the time lags of ozone response to GCR forcing (the bottom left panel in figure 4) shows that it responds without delay in the Northern Polar region and the Pacific coast of North America, while in other regions the lag is between 5-13 years.

The next link in our casual chain of relations, describing the GCR influence on the climatic variables, include the lower stratospheric ozone influence on the upper tropospheric water vapour. The upper right panel in figure 4 illustrates the synchronisation of the centennial variability of ozone density at 70 hPa and water vapour at 150 hPa. Note, that the regions of a strong ozone-water vapour coupling pretty well corresponds to the regions of the GCR influence on ozone. Particularly, the raise of the upper tropospheric humidity (especially well pronounced in the second half of 20-th century) corresponds to the ozone depletion in the tropical lower stratosphere (what is in line with our hypothesis). The positive water vapour response to the ozone abundance in some regions near 50^0N latitude seems to contradict to our hypothesis. However, the analysis of the time lags of humidity response shows that the raise of the water vapour density is observed with a delay of 9-11 years after the ozone enhancement (see bottom right panel in figure 4). Moreover, the direct comparison of the time series of both variables shows that they actually covariate in antiphase [38] – what is in accordance with our hypothesis.

6. Conclusions

This paper shows some more evidence for galactic cosmic rays' (GCR) influence on the near surface temperature and sea level pressure. The synchronisation between centennial variation of examined variables is significant at 95% level, which implies the existence of a causal links between them. The paper provides also a mechanism for GCR influence on the lower stratospheric ozone, which mediates the particles' influence down to the planetary surface. The ozone's impact on the near surface climatic variables is not direct, but is mediated by the changes in tropopause temperature and upper tropospheric humidity, which in turn affect the strength of greenhouse effect.

The estimation of the spatial distribution and the strength of the relations between different elements of the causal chain of links – describing the GCR impact on the atmospheric variables, supports the presented mechanism.

Acknowledgments

This work is partially supported by the Bulgarian FNI, contract DN 14/1 from 11.12.2017, and Contract KP-06-N34/1 from 30.09.2020. The authors are grateful to all providers of data, used in this study.

References

[1] IPCC, 2021, Climate Change 2021: The Physical Science Basis. Contribution of Working Group I to the Sixth Assessment Report of the Intergovernmental Panel on Climate Change [Masson-Delmotte, V., P. Zhai, A. Pirani et al. (eds.)]. Cambridge University Press 31 pp.

[2] Anderson TR, Hawkins E and Jones PD 2016 CO2, the greenhouse effect and global warming: from the pioneering work of Arrhenius and Callendar to today's Earth System Models *Endeavour* **40** 3 178-187 doi: 10.1016/j.endeavour.2016.07.002.

[3] Stips A, Macias D, Coughlan C et al. 2016 On the causal structure between CO2 and global temperature *Sci Rep* **6** 21691 https://doi.org/10.1038/srep21691.

[4] Svensmark H 2007 Cosmoclimatology: a new theory emerges *News Rev. Astron. Geophys*. **48** 1.18–1.24.

[5] Frigo E, Pacca I G, Pereira-Filho A J, Rampelloto P H and Rigozo N R 2013 Evidence forcosmic ray modulation in temperature records from the South Atlantic Magnetic Anomaly region *Ann. Geophys.* **31** 1833–41 https://doi.org/10.5194/angeo-31-1833-2013.

[6] El-Borie M A, Thabet A A, El-Mallah E S et al. 2020 Possible effects of galactic cosmic ray flux and low-cloud amounts on global surface temperature Pramana - *J Phys* **94** 45 https://doi.org/10.1007/s12043-019-1912-2.

[7] Solanki S K 2002 Solar variability and climate change: is there a link? *Astronomy & Geophysics* **43** 5 9-13.

[8] Wilcox J M 1975 Solar activity and the weather *J. Atmos. Sol.-Terr.Phy*. **37** 237–256.

[9] Kirkby J 2007 Cosmic Rays and Climate, *Surv. Geophys*. **28** 333–375.

[10] Solanki S K, Krivova N A and Haigh J D 2013 Solar Irradiance Variability and Climate *Annu. Rev. Astron. Astr*. **51** 311–351 https://doi.org/10.1146/annurev-astro-082812-141007.

[11] Dobrica V, Demetrescu C, Boroneant C, Maris G 2009 Solar and geomagnetic activity effects on climate at regional and global scales: case study—Romania *J Atmos Sol Terr Phys* **71** 17–18 1727–35.

[12] El-Borie M A, Abd El-Zaher M and El-Monier S 2012 Studying of the solar-climate interaction in Canada *World Environ* **2** 2 16–23.

[13] Olusegun C F, Rabiu A B, Ndeda J O H and Okogbue E C 2014 Trends of temperature and signature of solar activity in selected stations in Nigeria *Atmos Clim Sci* **4** 171–8.

[14] Frohlich C 2006 Solar irradiance variability since 1978 *Space Sci. Rev.* 125 53–65 https://doi.org/10.1007/s11214-006-9046-5.

[15] Foukal P, Frohlich C, Spruit H and Wigley T M L 2006 Variations in solar luminosity and their effect on the Earth's climate *Nature* **443** 161–166 https://doi.org/10.1038/nature05072.

[16] Georgieva K, Kirov B, Tonev P, Guineva V and Atanasov D 2007 Long-term variations in the correlation between NAO and solar activity: the importance of north south solar activity asymmetry for atmospheric circulation *Advances in Space Research* **40** 7 1152–66.

[17] Georgieva K, Kirov B, Koucka-Knızova P, Mosna Z, Kouba D and Asenovska Y 2012 Solar influences on atmospheric circulation *J.Atmos. Sol.terr. Phys*., 90–91, 15–25, doi.org/10.1016/j.jastp.2012.05.010.

[18] White W B, Lean J, Cayan D R and Dettinger M D 1997 Response of global upper ocean temperature to changing solar irradiance *Journal of Geophysical Research: Oceans* **102**, 3255–66 https://doi.org/10.1029/96JC03549.

[19] Barlyaeva T V 2013 External forcing on air–surface temperature: Geographical distribution of sensitive climate zones *Journal of Atmospheric and Solar-Terrestrial Physics* **94** 81–92 https://doi.org/10.1016/j.jastp.2012.12.014.

[20] Almeida A, Gusev A, Mello M G S, Martin I M, Pugacheva G, Pankov V M, Spjeldvik W N and Schuch N J 2004 Rainfall cycles with bidecadal periods in the Brazilian region *Geofísica Internacional* **43** 2 271-9.

[21] Peters D H W, Gabriel A and Entzian G 2008 Longitude-dependent decadal ozone changes and ozone trends in boreal winter months during 1960&2000 *Annales Geophysicae* **26** 1275–86 https://doi.org/10.5194/angeo-26-1275-2008.

[22] Kilifarska N A 2011 Long –term variations in the Stratospheric Winter Time Ozone Variability – 22 year cycle *Compt. rend. Acad. bulg. Sci*. **64** 6 867-874.

[23] Chapanov Y, Vondrák J and Ron C 2009 Common 22-year cycles of Earth rotation and solar activity, Proceedings of the International Astronomical Union **5** 407–9 https://doi.org/10.1017/S1743921309993000.

[24] Scafetta N 2010 Empirical evidence for a celestial origin of the climate oscillations and its implications *Journal of Atmospheric and Solar-Terrestrial Physics* **72** 951–970 https://doi.org/10.1016/j.jastp.2010.04.015.

[25] Laut P 2003 Solar activity and terrestrial climate: an analysis of some purported correlations *J Atmos Sol Terr Phys* **65** 801–812.

[26] Milankovitch M 1941, Kanon der Erdbestrahlung und seine Anwendung auf das Eiszeitenproblem. Académie royale serbe. Éditions speciales; 132 [vielm. 133]: XX, 633 S., Belgrad. (New English Translation, 1998, Canon of Insolation and the Ice-Age roblem. Pantic,

N. (Hrsg.), Beograd: Zavod Nastavna Sredstva, 634 s.)

[27] Velinov P I Y and Mateev L 2021 Modelling of the maximum spectrum of cosmic rays and their ionization during the minimums of solar cycles 23/24 and 24/25 *Compt. rend. Acad. bulg. Sci.* 74 12 1789-1798

[28] Pudovkin M I and Raspopov O M 1992 Physical mechanism of solar activity influence on the lower atmosphere and meteoparameters *Geomagnetism and Aeronomy* V. **32** n. 5 p. 3.

[29] Pudovkin M I and Babushkina S V 1991 The influence of electromagnetic and corpuscular radiation of solar flares on the intensity of zonal atmosphere circulation *Geomagnetism and Aeronomy* v. **31** n. 3 p. 493.

[30] Pudovkin M I andVeretenko S V 1992b Variations of meridional profile of atmospheric pressure during geomagnetic disturbances *Geomagnetism and Aeronomy* V. **32** n. 1 p. 118.

[31] Tinsley B A 1996 Solar wind modulation of the global electric circuit and apparent effects on cloud microphysics, latent heat release, and tropospheric dynamics *J. Geomag. Geoelec.* **48** 165–175 https://doi.org/10.5636/jgg.48.165.

[32] Svensmark H and Friis-Christensen E 1997 Variation of cosmic ray flux and global cloud coverage - A missing link in solar-climate relationships *Journal of Atmospheric and Solar-Terrestrial Physics* **59** 1225-32.

[33] Svensmark H 1998 Influence of cosmic rays on Earth's climate *Phys. Rev. Lett.* **81** 5027-30.

[34] Kenny D A 1979 *Correlation and causality* John Wiley & Sons Inc, New York.

[35] Bakhmutov V, Yaremenko L, Melnyk G and Shenderovska O 2006 Magnetic field of Antarctic according to modelling calculations IGRF and supervision of observatories *Ukr. Antarct. J.* 4–5 64–71 (in Russian).

[36] Melnyk G, Bakhmutov V. and Shenderovska O 2014 Antarctic geomagnetic field changes in the last century *Ukr. Antarct. J.* **13** 75–80 (in Russian).

[37] Kilifarska N 2015 Bi-decadal solar influence on climate, mediated by near tropopause ozone *J. Atmos. Solar-Terr. Phys.* **136** 216–230.

[38] Kilifarska N 2017 Hemispherical asymmetry of the lower stratospheric O_3 response to galactic cosmic rays forcing *ACS Earth & Space Chem.* **1** 80–88.

[39] Tassev Y, Velinov P I Y, Mateev L and D Tomova 2003 Comparison between effects of solar proton events and geomagnetic storms on the ozone profiles *Adv. Space Res.* **31** 9 2163–68.

[40] Velinov P I Y 2006 Cosmic ray influence on the system ionosphere-atmosphere through ionization, chemical and electrodynamic processes. CR as key governing the Sun-Earth connections, Rep. Int. Symp. Recent Observations and Simulations of the Sun-Earth System (ISROSES), Varna, 17–22 September 2006, *Progr. Abstr. Book*, p. 7, p. 103, Sofia, Heron Press, Ltd. 1–33.

[41] Kilifarska N A 2013 An autocatalytic cycle for ozone production in the lower stratosphere initiated by Galactic Cosmic rays *Compt. rend. Acad. bulg. Sci.* **66** 2 243-252

[42] Young J A 2003 Static stability In: *Encycl. Atm. Sci.* (eds J. R. Holton, J. A.Curry, J. A. Pyle) London, Academic Press **5** 2114–20.

[43] North G R and Eruhimova T L 2009 Atmospheric Thermodynamics: Elementary Physics and Chemistry Cambridge University Press.

[44] Inamdar A K, Ramanathan V and Loeb N G 2004 Satellite observations of the water vapor greenhouse effect and column longwave cooling rates: relative roles of the continuum and vibration-rotation to pure rotation bands *J. Geophys. Res. Atmos.* 109 https://doi.org/10.1029/2003JD003980.

[45] Kilifarska N A, Bakhmutov V G and Malnyk G V 2020 The Hidden Link Between Earth's Magnetic Field and Climate, 1st Edition *Elsevier* 218 pp.

Geomagnetic control of cosmic radiation and its impact on the irregularly distributed lower stratospheric ozone

N A Kilifarska[1], **K Peqini**[2]

[1] Climate, Atmosphere and Water Research Institute, Bulgarian Academy of Sciences, 66 Tsarigradsko Shosse Blvd, 1784 Sofia, Bulgaria

[2] Department of Physics, Faculty of Natural Sciences, University of Tirana, Blvd. Zogu I, Nr. 25, Tirana, Albania

natalya_kilifarska@yahoo.co.uk

Abstract. This article has two main objectives: (i) to elucidate the impact of non-dipolar part of geomagnetic field on the spatial distribution of cosmic radiation received over the globe, and (ii) to establish the influence of particles (with different origin and energies) on the atmospheric ozone profile. Our analyses show more evidence that lower stratospheric ozone at subtropics is sensitive to the radiation trapped by geomagnetic filed in Van-Allen radiation belts. At the same time, the high latitude ozone is vulnerable to the solar plasma, hitting the Earth after solar proton events.

1. Introduction

The effect of energetic particles on atmospheric chemistry is widely exploited in scientific research. The common expectation is that strongest particles' impact should be expected in polar regions, due to the strongest flux of precipitating particles. This presumption is not fully correct, because the particles' impact on atmospheric chemistry depend on their energy, and respectively – on the depth of their penetration in the atmosphere. For example, the low energy particles, which are substantial part of the particles' flux reaching the polar regions, are absorbed in the upper atmosphere [1-4]. Consequently, if one is looking at the radiation's impact on the lower stratosphere, the effect will be minimal. On the other side, the geomagnetic shielding of the lower latitudes filters the low energy particles. Thus, the tropical and extratropical regions of the Earth are reachable by particles, which are able to overcome the geomagnetic defence [5]. In addition, these latitudes are influenced by particles confined in the Earth's radiation belts, which are subject of geomagnetic lensing in a gradient geomagnetic field.

Due to the altitude dependence of atmospheric density, humidity and compositional structure, the particles absorbed at different atmospheric level activate different ion-molecular reactions – having, in turn, different impact on the atmospheric chemistry. For example, low energy particles absorbed at mesospheric levels, activate production of HO_x (OH and HO_2) and NO_x (NO, NO_2, and NO_3) compounds, leading to the destruction of ozone at these levels [6-9]. On the other hand, the secondary ionisation, produced by galactic cosmic rays in the lower stratosphere, activates ion-molecular reactions producing ozone [10].

This paper is aimed to make a brief resume of the physical basis of particles' movement in gradient magnetic field, including the processes of energy transfer to the atmosphere, as well as of the effect of geomagnetic inhomogeneity on the spatial distribution of particles flux over the world. In addition, the paper shows an evidence for ozone sensitivity to the short lasting decreases of galactic cosmic ray flux (Forbush decrease), depending on the energy of precipitating particles.

1.1. Charged particles movement in a dipolar geomagnetic field

Cosmic radiation, consisting of charged particles (mainly protons and electrons and a small amount of heavier ions and neutrons) continuously penetrate in Earths' atmosphere. Some of the particles originate from the solar atmosphere; others have galactic and extragalactic origin. Those of particles, which enter the magnetosphere along the open magnetic field lines are either repelled back in space by geomagnetic field, or are lost in the atmosphere, due to the multiple collisions with the atmospheric molecules.

Particles approaching the magnetosphere in a region of closed magnetic field lines become trapped or quasi-trapped in the geomagnetic field, if their energy is insufficient to overcome the geomagnetic cutoff rigidity. They are urged by the Lorentz force (refer to equation 1) to move along the magnetic field lines on spiral trajectories (the result of a combined *circular* and a *field align* motion), continuously bouncing between the Northern and the Southern Hemispheres.

$$m\frac{dv}{dt} = q[\mathbf{E} + (\mathbf{v} \times \mathbf{B})]; \quad \mathbf{v} = \frac{dr}{dt} \tag{1}$$

where: **B**(r,t) and **E**(r,t) are external magnetic and electric fields – function of the spatial dimensions and time, **r** and **v** are respectively particle radius vector and velocity; m is particle mass and q – its charge.

The radius of the particles *circular* trajectory, known as a *Larmor* radius (refer to equation 2), is determined by: (i) the projection of particle's velocity on a direction perpendicular to B, i.e. $v_{\perp 0}$; (ii) the particle's charge q and mass m; and (iii) the magnetic field strength **B**, [11], i.e.:

$$r_B = \frac{m \cdot v_{\perp 0}}{q \cdot B} \tag{2}$$

Formula (2) clearly shows that the electrons and protons rotate in opposite directions (by convention the positive, clockwise, is the angular velocity of electrons, when B is away from the observer).

The field-aligned motion v_{II} is defined by the projection of initial particle's velocity along the geomagnetic field line. In a gradient magnetic field, like geomagnetic field, the v_{II} depends on the geomagnetic gradient along the magnetic field line, [11], i.e.:

$$v_{II}(t) = v_{II_0} - \frac{v_\perp^2}{2 \cdot B_z}\left(\frac{\partial B_z}{\partial z}\right) \cdot t \tag{3}$$

Equation (3) shows that when a particle enters the region of *increasing* field intensity its velocity, which is parallel to the magnetic field, is retarded and at some level it becomes zero. The further raise of magnetic gradient changes the sign of particle's velocity, i.e. it is reflected backward. The region where charged particles are subject to a reflection is called a *magnetic mirror*.

Besides the helical movement of particles along geomagnetic field lines, they also perform additional movement in a direction perpendicular to the magnetic field lines – known as magnetic drift. This type of motion is determined by the non-uniformity of Earth's magnetic field in the direction perpendicular to ***B***, and by the magnetic field's curvature (sees equation 4):

$$v_{drift} = \frac{m}{q \cdot B^2}\left(v_\perp^2 \cdot \frac{\mathbf{B} \times \nabla \mathbf{B}}{2\mathbf{B}} + v_{II}^2 \frac{\boldsymbol{\rho} \times \mathbf{B}}{\rho^2}\right) \tag{4}$$

where **B** is the magnetic vector, **ρ** – the radius of curvature of the geomagnetic field lines, v_{II} and $v_\perp$ are projections of particles velocities parallel and perpendicular to geomagnetic field line; **q** and **m** are respectively particles charge and mass. The first term in the brackets corresponds to the magnetic *gradient* perpendicular to the field lines, while the second term – to their *curvature*.

Formula (4) shows also that particles drift across the magnetic field lines depends on their charge **q**, and consequently leads to charge separation, which generates electric field **E** along the drift direction. The combined effect of **E** and **B** fields induces an **E**x**B**/B^2 drift of particles, which displaces positive ions and negative electron in the same direction, perpendicular simultaneously to **B** and to **E**. These charged particles are then "lost" in the ambient atmosphere, where they release their energy, producing showers of secondary particles.

1.2. Particles' energy released in the atmosphere

The different radiations behave differently in their passage through the matter. Heavy particles, for example, lose little of their energy in a single collision, so in fact they are never deflected (except by a rare encounter with a nucleus). Thus, the heavy particles lose their energy primarily through Coulomb scattering on the outer electron shells, delivering small amounts of energy to the orbital electrons and leaving behind excited molecules or electron-ion pairs [12]. Light particles (i.e. electrons or positrons) can lose a large fraction of their energy in a single collision with an electron. In fact, they can lose all of it, creating a knock-on electron of about the same energy. Light particles are much more easily deflected, so their paths are not as straight as those of heavy particles.

The relatively low energetic particles contribute to the energy loss mainly through the ionization of atoms and molecules filling the medium, while the very energetic ones (with energies greater than 1 GeV) contribute with nuclear interactions with atomic nuclei of the atmosphere. The energy loss of energetic particles moving through matter is described by Bohr-Bethe-Bloch formula, having the following form:

$$-\left[\frac{dE}{dx}\right] = \frac{4\pi n z^2}{m_e v^2} \cdot \left(\frac{e^2}{4\pi\varepsilon_0}\right)^2 \cdot \left[\ln\left(\frac{2 m_e v^2}{I(1-\beta^2)}\right) - \beta^2\right] \tag{5}$$

where z are charges of incident particles; e – electron charge; n – electron density of the target; ε_0 the vacuum permittivity; I – mean excitation energy of the target atom (molecule); m_e – mass of electron, v and c – are speeds of incident particle and speed of light, and $\beta = v/c$.

The equation (5) shows that energy lost by a particle in the atmosphere depends inversely on the square of its speed, $1/v^2$. This indicates that the faster the particle is the less energy it transfers to the molecules of the medium. Consequently, in regions with higher geomagnetic gradient, where according to equation (3) particles velocity slow down, the energy lost on ionisation of atmospheric molecules must increase.

1.3. Cosmogenic isotopes in Earth's atmosphere

As mentioned above, cosmic ray particles, with energies greater than 1 GeV, are able to interact directly with the nucleus of atmospheric atoms. In these inelastic collisions a large part of the energy of hitting protons is carried away by protons, neutrons and alpha particles expelled from the target nucleus. Furthermore, these secondary particles develop hadronic cascades in the atmosphere producing a bulk of new particles like *kaons, pions, muons* etc. Some of the energy of cosmic ray protons is transferred to the target nuclei. The exited nucleus can break up on fragments and this process is called *spallation*. The large variety of cosmogenic radionuclides has been created by the spallation, and all of them have masses smaller than that of the target nucleus. Moreover, the exact structure of target nucleus is only of minor importance [13]. For example, beryllium isotope ^{10}Be can be produced by four different reactions:

$$^{14}N + n \rightarrow 2n + 3p + {}^{10}Be \tag{6}$$

$$^{14}N + p \rightarrow n + 4p + {}^{10}Be \tag{7}$$

$$^{16}O + n \rightarrow 3n + 4p + {}^{10}Be \tag{8}$$

$$^{16}N + p \rightarrow 2n + 5p + {}^{10}Be \tag{9}$$

where p and n designate proton and neutron. The thresholds energy for ^{10}Be production from nitrogen is 15 MeV for neutrons, and 33 MeV for protons, while for production from oxygen – it is about 30 MeV for both nucleons. The efficiency of radionuclides production is determined mainly by the *hardness* of the differential energy spectra of protons and neutrons. Analyses of Beer et al. [13] reveal that the *neutron–nitrogen* interactions have the largest impact in the ^{10}Be production.

The neutrons are created mainly by hadronic cascades, because galactic cosmic rays entering the atmosphere do not normally contain neutrons (except those in the nuclei of helium and heavier elements). The higher efficiency of neutrons in the process of spallation, compared to the energetic protons, is due to the fact that they do not need to overcome the Coulomb barrier of the target atom, experienced by the protons. The difference between proton and neutron becomes especially obvious at low energies, because the positively charged proton is repelled by the protons in the target nucleus (Coulomb barrier), while the neutron can easily enter the nucleus at very low energies.

An illustration of the higher impact of neutrons in production of cosmogenic isotopes is the fact that ~99% of the ^{14}C isotope is produced by thermal neutrons with a typical energy of 0.025 eV [13], through the reaction:

$$^{14}N + n \rightarrow p + {}^{14}C \qquad (10)$$

2. Heterogeneous geomagnetic field and its impact on particles movements

The non-dipolar part of geomagnetic field becomes more apparent close to the planetary surface. The geomagnetic irregularities affect particles' motion and these disturbances are discussed in the following subsection.

2.1. Geomagnetic lensing of charged particles

The geomagnetic field is a vector sum of a magnetic dipolar field (the dominant part), non-dipole field related to the heterogeneous structure of the deep Earth's interior, magnetic properties of the crustal rocks and magnetic field of external sources. The spatial heterogenity of geomagnetic field is well visible even in its main part – the core magnetic field, which is induced in the liquid upper Earth's core. The bi-directional geomagnetic gradient (i.e. *latitudinal* – along the meridian and *longitudinal* – along the given parallel) is well visible in figure 1. The figure presents the radial component (B_r – directed downward), the polar one (B_S – directed southward) and azimuthal component (B_E – directed eastward) of the core geomagnetic field (the left column in figure 1) as calculated by the CHAOS5 magnetic model, for 2010. The right column shows the changes in these three components, experienced by charged energetic particles in their movement from ionospheric levels to the Earth's surface.

Under the influence of bi-directional geomagnetic gradient, the particles with different sign are deflected in different direction. Note that at the lowest part of their circular trajectories, the protons enter Earth's atmosphere from the west, while electrons – from the east. Consequently, in the x-y plane (with x directed to the east, and y – to the north), the protons are shifted south-westward in regions with a positive cross-longitudinal gradient and south-eastward – in regions with a negative gradient (refer to equation 4 and figure 1). Consequently, the overall *westward* drift (forced by the magnetic curvature and cross-latitudinal gradient) is reduced by the *eastward* component – exerted in regions with negative azimuthal magnetic gradient. Furthermore, the electric field (induced by the charge separation of impending particles) is significantly reduced in these regions. Finally, the number of particles expelled outside the magnetic trap (due to the $(ExB)/B^2$ electric drift) is much less. More precisely, only a few of them have a "chance" to be lost in the atmosphere in said regions. As a result, cosmic radiation – detected by the ground based neutron monitors – is significantly diminished.

Oppositely, in regions with positive azimuthal geomagnetic gradients, the south-westward drift component changes slightly the direction, but not the amplitude of the westward drift, impelled by the magnetic curvature and latitudinal gradient. Consequently, in these regions the induced electric field – resulted from the charge separation of arriving particles – is much stronger. It will intensively expel the charged particles outside the magnetic trap through the imposed $ExB)/B^2$ drift. Furthermore, these

Journal of Physics: Conference Series **2255** (2022) 012013 doi:10.1088/1742-6596/2255/1/012013

particles interact with the atmospheric molecules creating secondary electrons, ions and nuclear products, giving rise to the ionisation of the lower atmosphere, and to the radiation measured by the ground based neutron monitors.

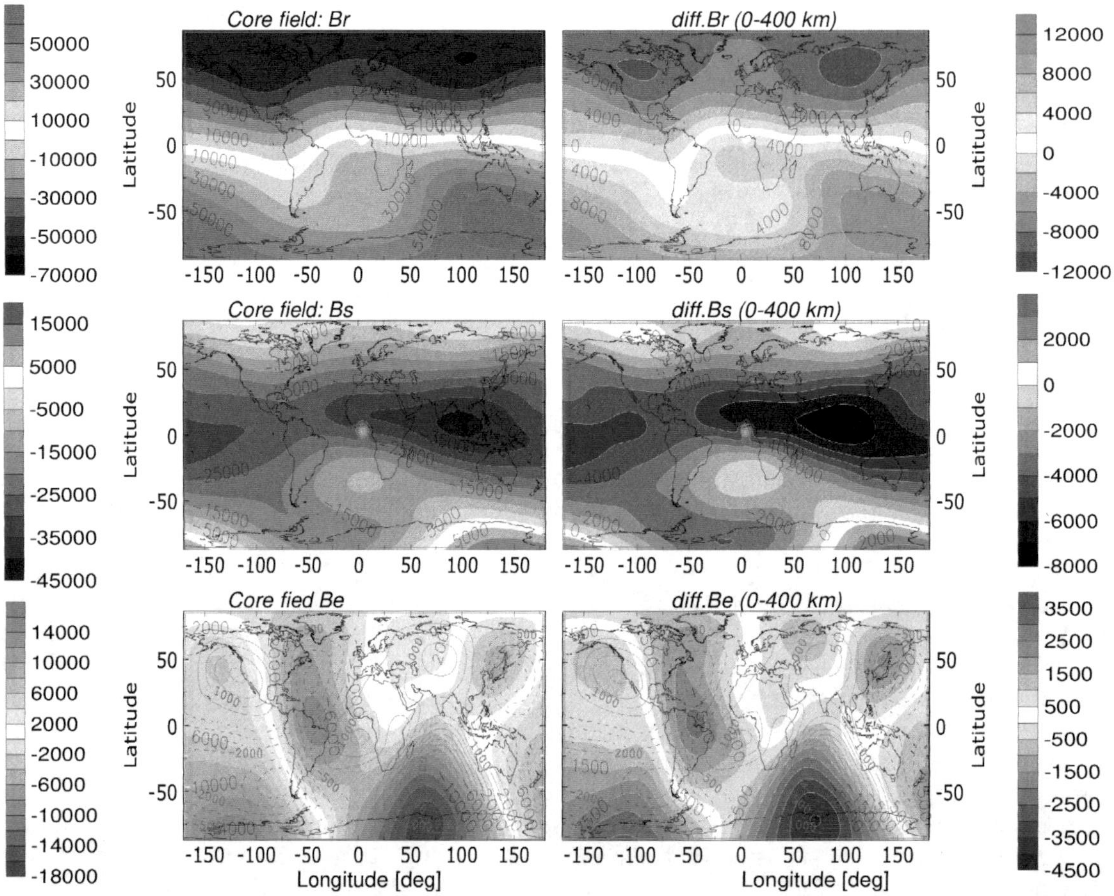

Figure 1. (left) Spatial distribution of radial (B_r), polar (B_S) and azimuthal (B_E) component of the Earth's core magnetic field, calculated from the CHAOS5 [14] model, for 2010; (right) Difference of geomagnetic strength at the surface and at 400 km above it. The overdrawn on the azimuthal component B_E is the geomagnetic longitudinal gradient (the bottom panes).

2.2. Hemispherical asymmetry of geomagnetic field and it impact on the trapped particles' precipitation

The number of very high energy particles entering Earth's atmosphere is relatively small and their effect is negligible [15,16], because the Bohr-Bethe-Bloch formula (5) shows that energy released in the atmosphere is inversely proportional to the particles speed [17,18]. The less energetic particles are attracted by geomagnetic field and are compelled by the Lorentz force to move around and along the magnetic field lines on spiral trajectories. The confinement of any particle in the gradient magnetic field $\mathbf{B}$ depends on the ratio between the maximum field strength $\mathbf{B}_{\mathbf{max}}$ in the polar regions (where the backward reflection of trapped particles occurs) and the equatorial magnetic field strength $\mathbf{B}_0$, i.e.

$$\sin(\alpha) = \sin(\alpha_0) \cdot \left(\frac{B_{max}}{B_0}\right)^{-1/2} \tag{11}$$

where α_0 is the angle between velocity vector of the arriving particle and the equatorial magnetic field – known as an *equatorial pitch angle*, and $\boldsymbol{\alpha}$ is the continuously changing pitch angle (with particle motion along the magnetic field line).

Any particle is assumed trapped by the magnetic field, when the angle $\boldsymbol{\alpha}$ becomes greater than $\frac{\pi}{2}$, because at this point – known as *magnetic mirror* – the particle reverses its direction of movement, remaining confined by the magnetic field line. Formula (11) shows that particles approaching Earth's magnetosphere at very small angles could not exceed the pith angle $\frac{\pi}{2}$, and when enter the mirror point these particles are "lost" in the atmosphere. The minimum value of angle α_{0m} (for which the maximum magnetic field is still able to reflect particles) is called *loss cone*. If a particle arrives at angle lower than solid angle defined by α_{0m}, it will be lost in the ambient atmosphere on its motion along the magnetic field line. Formula (11) shows also that the efficiency of magnetic mirror to reflect charged particles does not depend neither on particles speed, no on their charge and mass.

The geomagnetic field near the poles is stronger in the Southern Hemisphere, compared to those in Northern Hemisphere. Consequently, in the case of isotropic particles' flux arriving at the magnetopause almost every third particle will be confined in the Southern Hemisphere, while in the Northern Hemisphere less than ¼ of all arriving particles are trapped – due to its larger *loss cone*. This means that some of particles confined in the Southern Hemisphere could not be held by the weaker geomagnetic field in the Northern Hemisphere. The expected result is more particles precipitating in the Northern Hemisphere.

Energetic particles penetrating deeper in the atmosphere creates showers of secondary particles, produced from their interaction with atmospheric molecules – deeper the penetration, wider the showers. In the lower stratosphere, the number of secondary products dramatically increases, becoming maximal at a certain level. This level is known as the *Regener-Pfotzer maximum*. Beneath it, the concentration of secondary ions and electrons decreases again.

3. Ozone response to the short lasting Forbush decreases

Analyses of the centennial variations of galactic cosmic rays (GCR) and ozone at 70 hPa reveals that the hemispherical and longitudinal asymmetry of the latter could be attributed to the energetic particles' impact, the intensity and spatial distribution of which is modulated by the heterogeneous geomagnetic field [19,20]. The mechanism of particles' influence on the ozone density is related to the activation of ion-molecular reactions leading to a production (or destruction) of ozone [20].

On the other hand, the GCR flux, arriving at the Earth's magnetopause, is preliminary modulated by the heterogeneous heliomagnetic field. The 11-year periodicity of solar dynamo, as well as the emission of solar plasma or electromagnetic radiation from the Sun, is the main source of magnetic irregularities in heliosphere. The clouds of high speed and dense solar plasma reflect the galactic cosmic rays (originating from outside the solar system). At the Earth's surface, this process is registered as a sudden decrease of GCR flux (known as Forbush decrease). We have examined the consequences of the coronal mass ejection, which initiated one of the strongest geomagnetic storm during the 24-th solar cycle, in March 2015, known as the St. Patrick's Day storm.

The peculiarity of this storm is that Forbush decrease (FD) does not occur simultaneously over the world, what usually happened when a relatively weaker interplanetary magnetic cloud strikes the dusk side of the magnetosphere [21]. We have exploited this situation to investigate the local response of ozone profile to the short lasting decrease of the cosmic ray flux reaching the surface.

Figure 2 illustrates the time of appearance of FD in two neutron monitors having similar longitudes, but different latitudes, and respectively different geomagnetic cutoff rigidity. This means that ozone profiles over these observatories is affected by particles with different energies and different origin. At the same time, the two monitors are placed near 30^0E longitude, which excludes

Journal of Physics: Conference Series **2255** (2022) 012013 doi:10.1088/1742-6596/2255/1/012013

the potential influence of the geomagnetic field longitudinal gradient. Moreover, the FD onset in both observatories is shifted by four days, which allows us to analyse the ozone's response to the global (storm related) and local changes of the atmospheric ionization – associated to the Forbush decrease.

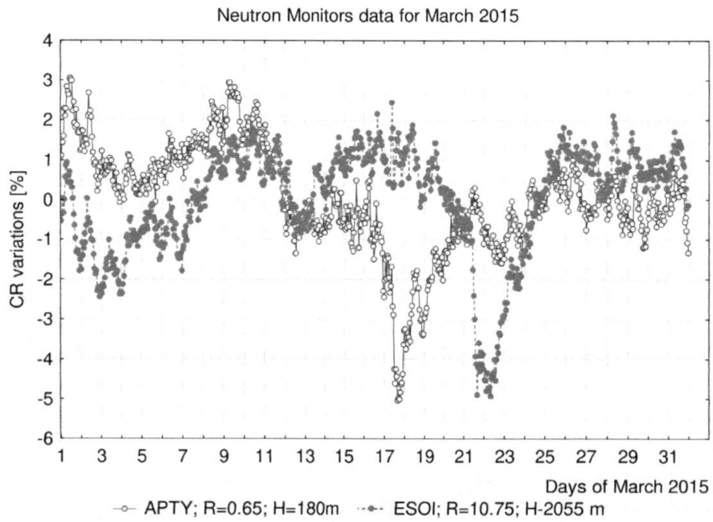

Figure 2. Time series of cosmic radiation measured in two neutron monitors' locations – Apatite (APTY; R=0.65, Alt=180m) and Emilio Segre Observatory in Israel (ESOI; R=10.75, Alt=2055 m). Note the time lag of ~4 days between Forbush decrease observed in APTY and ESOI.

Figure 3 presents the response of ozone profile to the variations in cosmic rays arriving at Apatite neutron monitor. Note that the Forbush decrease commencement is followed by a reduction of ozone density in the upper stratosphere, and its pile up in the lower stratosphere – starting at March 18, 2015 (the red curve in figure 3, left).

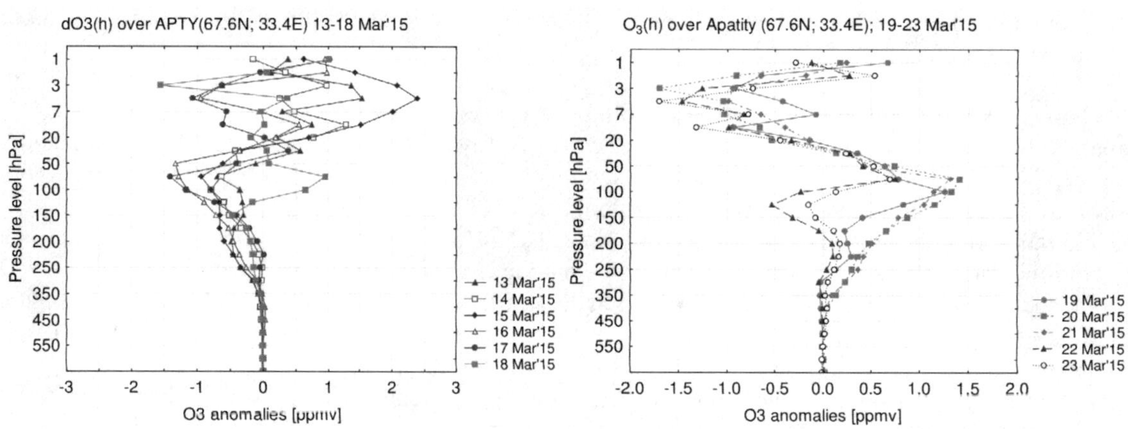

Figure 3. Ozone profiles over Apatite before, during and after the geomagnetic storm on 17-18 March 2015.

The low magnetic rigidity of Apatite determines the free access of energetic particles, arriving along the open geomagnetic field lines, to the polar stratosphere. When the interplanetary shock-wave passes the Earth, the solar particles reaching the upper stratosphere activate ozone destructive chemical reactions there. The reduced optical thickness of the upper stratospheric ozone, in turn, allows penetration of solar ultraviolet radiation deeper in the lower stratosphere, where begins the process of ozone self-restoration [22]. This transformation of ozone profile is well visible during the four days following the onset of geomagnetic storm and FD over Apatite, i.e. within the period 18-21 March 2015 (refer to figure 3).

On the other hand, the ESOI site is assessable mainly to the particles trapped in the Earth's radiation belts. This explains the smallest O_3 response over ESOI observatory to the global geomagnetic storm of March 17-18, 2015 (see figure 4). The weak positive anomalies in ozone mixing ratio, between 50 and 200 hPa (the red curves in figure 4), starts at March 17 and continues until the onset of FD in March 21, 2015. This enhancement of the lower stratospheric ozone is accompanied by its depletion at the upper stratospheric levels. The latter fact is an indication of the self-restoration mechanism [22] standing behind the positive ozone anomalies in the lower stratosphere within the period 17-20 March.

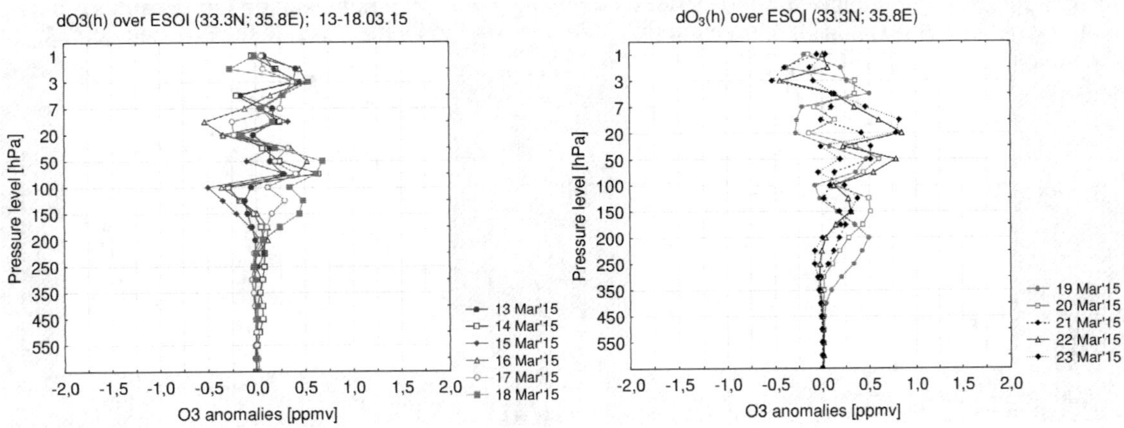

Figure 4. Ozone profile over Emilio Segre Observatory and its variability during geomagnetic storm in 17-18 March 2015.

With other words, the mid-latitude ozone response to the geomagnetic storm is negligible, due to the magnetic shielding of middle latitudes, preventing the penetration of lower energy solar particles at these latitudes. Let see what is the response of the ozone's profile to the sudden decrease of GCR intensity (i.e. Forbush decrease), which has occurred over the Emilio Segre Observatory at 21-22 March, 2015. The temporal evolution of daily ozone profiles is presented in figure 5. Note the strong enhancement of O_3 density in the middle stratosphere (7-30 hPa), following the recovery of the cosmic radiation after the Forbush decrease (since 24 March 2015). A short-lasting weaker peak is visible also near 100 hPa. In this case, the raise of the lower stratospheric ozone density could not be attributed to the self-restoration mechanism, because the ozone density was enhanced also in the upper stratosphere. Consequently, the most probable explanation of the lower stratospheric ozone abundance is the activation of the ion-molecular reactions producing ozone [18,19].

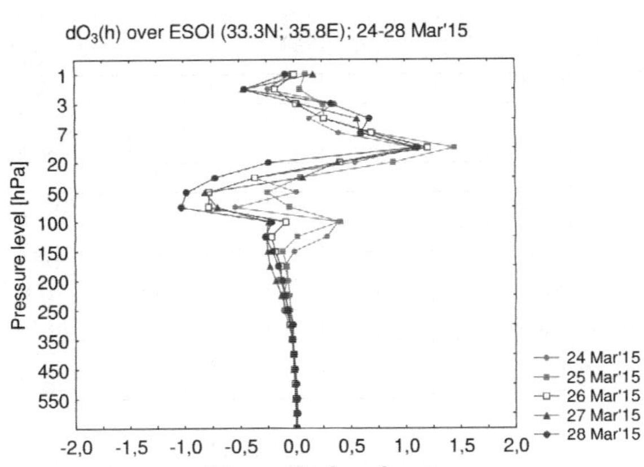

Figure 5. Response of ozone profile over Emilio Segre Observatory to the Forbush decrease in March 21-22, 2015

The quantitative estimation of the synchronicity between ozone and GCR temporal variability is performed by the use of two methods – the lagged cross-correlation analysis and the artificial neural network method. The results are presented in figure 6. The cross-correlation analysis reveals the in-phase synchronisation between GRC and ozone variability in the upper stratosphere over the Apatite (red contour in figure 6A). The lower stratospheric O_3 over Apatite is moderately correlated with cosmic radiation, but with opposite sign. The strength of CR-ozone relation over Emilio Segre Observatory is weaker and more complicated. Correlation coefficients above 0.5 are found in the upper stratosphere, as well as at 50-70 hPa. Unlike the Apatite, where the tropospheric O_3 does not correlate with cosmic radiation, the temporal variations of both variables over Israel are moderately synchronised, and their correlation coefficients are between 0.4 and 0.5. The all correlation coefficients are statistically significant at 0.05 level, meaning that probability of making a type I error (i.e. erroneous rejection of null hypothesis for non-correlating variables) do not exceeds 5%.

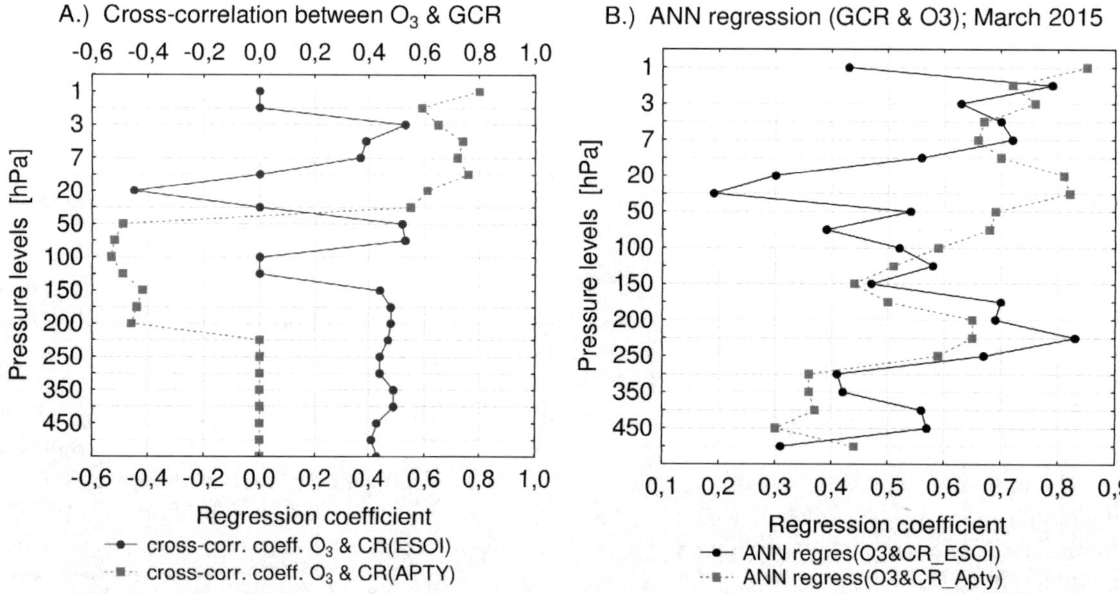

Figure 6. (A.) Cross-correlation between galactic cosmic rays measured in Apatite (red contour) and Emilio Segre Observatory (blue contour), and corresponding ozone profiles over both sites; (B.) Similarity in temporal evolution of O_3 density and cosmic radiation measured at Apatite (red contour) and Emilio Segre Observatory (black contours) during the March 2015.

The results derived with neural network are presented in figure 6B. High degree of similarity between GCR and ozone temporal variations has been found in the upper stratosphere over both observatories. In Apatite two more peaks of correspondence are visible – at 30 and 225 hPa. In ESOI observatory the highest ever similarity between cosmic radiation and O_3 mixing ratio is found between 175 and 225 hPa.

Comparison of the results derived by the two methods shows that cross-correlation technics give only a rough estimation of the possible relations between analysed variables. The neural network technic is more flexible, allowing better description of height variable relation between cosmic radiation and ozone. However, it has to be bear in mind that the anti-phase covariance is interpreted by this method as dissimilarity.

4. Discussion and Conclusions

The influence of energetic particles on the atmospheric composition is in the core of many hypotheses attempting to explain the imprint of solar variability on climate. However, the physical basis of the processes controlling the intensity of particles' flux, their depth of penetration in the atmosphere (together with the altitude dependence of the chemical reactions activated there), the spatial distribution of particles' precipitation over the globe, etc., are not well understood by the climatic community. So, one of the aims of this paper was to summarise the most important of the recent achievements, which takes into account the heterogeneous character of geomagnetic spatial distribution and its effect on the particles movement in Earth's atmosphere.

The second purpose of this article .was to elucidate the response of atmospheric ozone to the changes in galactic cosmic rays' intensity, over two ground based neutron monitors with different geomagnetic cutoff rigidity along the 30^0 E meridian. Such a treatment of the problem allows establishment of differences in atmospheric response to the impact of particles with different origin and energies. To achieve this purpose, we have select for analysis the solar coronal mass ejection on 15 March 2015, which produces the strongest, during the solar cycle 24, geomagnetic storm on 17 March 2015. One of the specific characteristic of this storm is the non-uniform appearance over the world of the GCR intensity sudden decrease (known as Forbush decrease, FD). This fact allows us to investigate the response of ozone profile not only to geomagnetic storm, itself, but also to the FD.

We show that the onset of Forbush decrease (related to the solar coronal mass ejection on 15 March 2015) is followed by positive anomalies in the lower stratospheric ozone, detected in both analysed observatories – Apatite (67.57^0 N; 33.39^0 E) and Emilio Segre Observatory (33.3^0 N; 35.8^0 E). The mechanisms of ozone enhancement in both sites, however, are different. Thus at Apatite, which is accessible for particles arriving along the open geomagnetic field lines, the abundance in the lower stratospheric ozone density is due to the process of ozone self-restoration, after its destruction at higher levels by the lower energy solar particles, arriving with interplanetary magnetic shock [11].

On the other hand, the subtropical Emilio Segre Observatory in Israel is influenced mainly by the energetic particles trapped in Earth's radiation belts. For this reason, the onset of Forbush decrease in arriving cosmic radiation flux is followed by a depletion of the lower stratospheric ozone density – due to the reduced amount of available atmospheric ionisation and reduced rate of ozone production. In the period of recovery after the Forbush decrease, however, a pile up of ozone density is found – at lower and in the upper stratospheric levels. While the upper stratospheric effect (at 5-10 hPa) could be partially attributed to the transition from winter to the summer raise of ozone density, the enhancement in the lower stratospheric ozone should be undoubtedly assigned to the raise of the atmospheric ionisation after the Forbush decrease, and activation of additional ozone production near the Regener-Pfotzer maximum [19,20].

In resume, the atmospheric ozone profile is very sensitive not only to the solar particles related to the geo-effective solar eruptions (broadly elucidated in scientific literature), but also to the galactic cosmic rays, and more specifically to the secondary ionisation produced by them in the lower stratosphere. The latitude dependence of the ozone response is due to the different energies of the impacting particles. The lower stratospheric O_3 at latitudes shielded by the closed geomagnetic filed lines is sensitive to the radiation trapped by geomagnetic filed in Van-Allen radiation belts. At the same time, the high latitude ozone is vulnerable to the solar plasma, hitting the Earth after solar proton events.

5. Data and methods used

The hourly values of neutron monitors counting rates have been derived by the freely available data at Neutron Monitors Data Base portal (http://www01.nmdb.eu), and the IZMIRAN data server (http://cr0.izmiran.ru/common/links.htm). Provided data are pressure and efficiency corrected. The relative changes in cosmic radiation in [%] are calculated as deviations from the monthly mean values, normalised by the same mean and multiplied by 100.

Gridded data for atmospheric ozone profile has been taken from the ERA-Interim reanalysis (http://apps.ecmwf.int/datasets/data/interim-full-daily/levtype=sfc/). The deviations from the monthly mean values have been used for assessment of the short-term fluctuations in ozone's profile.

Two methods have been used as quantitative estimators of similarity in temporal variability of cosmic radiation and ozone profiles – the classical cross-correlation method [23] and the artificial neural network one [24]. Statistical significance of correlation coefficients, derived by the cross-correlation analysis, has been estimated through applying Student's t-test, with a threshold for α=0.05.

The most suitable model (among the bulk of possibilities, provided by the ANN) is chosen through optimisation between the following criteria: (i) maximisation of model's performance (regression coefficients) derived for each of training and testing subsamples, and (ii) minimisation of model's error of both samples.

Acknowledgments
This work is supported by the Bulgarian FNI, contract DN 14/1 from 11.12.2017, and Contract KP-06-N34/1 from 30.09.2020. We acknowledge the NMDB database (www.nmdb.eu), founded under the European Union's FP7 program (contract no. 213007), as well as the IZMIRAN neutron monitors database (http://cr0.izmiran.ru/common/links.htm) for the data provided. Special thanks also to the project teams of ERA Interim reanalysis, providing gridded data for meteorological variables.

References

[1] Nicolet, M., 1975. On the production of nitric oxide by cosmic rays in the mesosphere and stratosphere, Planetary and Space Science 23(4), 637–649.
[2] Thorne, R.M., 1980. The importance of energetic precipitation on the chemical composition of the middle atmosphere, Pure Appl. Geophys. 118, 128–151.
[3] Vampola, A.L., Gorney, D.J., 1983. Electron Energy Deposition in the Middle Atmosphere, J.Geophys.Res. 88(A8), 6267-6274.
[4] Callis, L.B., Blake, D.N., Baker, J.B., Lambeth, J.D., Boughner, R.E., Natarajan, M., Klebesadel, R.W., Gorney, D.J., 1991. Precipitating relativistic electrons: their long-term effects on stratospheric odd nitrogen levels, J. Geophys. Res 96(2), 2939–2976.
[5] Shea, M.A., Smart D.F., 2004. Preliminary study of cosmic rays, geomagnetic field changes and possible climate changes, Adv. Space Res. 34, 420–425.
[6] Jackman, C.H., Frederick, J.E., Stolarski, R.S., 1980. Production of Odd Nitrogen in the Stratosphere and Mesosphere: An Inter-comparison of Source Strengths, J.Geophys.Res. 85(C12), 7495-7505.
[7] Rusch, D.W., Gerard, J.C., Solomon, S., Crutzen, P.J., Reid, G.C., 1981. The effect of particle precipitation events on the neutral and ion chemistry of the middle atmosphere – I. Odd nitrogen, Planet, Space Sci. 29(7), 767-774.
[8] Hauchecorne, A., Bertaux, J.-L., Lallement, R. , 2006. Impact of solar activity on stratospheric ozone and NO2 observed by GOMOS/ENVISAT, Space Sci. Rev. 125, 393–402.
[9] Seppälä, A., Clilverd, M.A., Rodger, C.J., Verronen, P.T., Turunen, E., 2008. The effects of hard-spectra solar proton events on the middle atmosphere, J. Geophys. Res. 113, A11311, doi:10.1029/2008JA013517.
[10] Kilifarska N.A. An autocatalytic cycle for ozone production in the lower stratosphere initiated by Galactic Cosmic rays, Compt. rend. Acad. bulg. Sci., 66(2), 243-252, 2013.
[11] Moisan, M., Pelletier, J., 2012. Individual Motion of a Charged Particle in Electric and Magnetic Fields. In: Physics of Collisional Plasmas. Springer, Dordrecht.
[12] Wilson, J. W., Townsend, L. W., Schimmerling, W., Khandelwal, G. S., Khan, F., Nealy, J. E., Cucinotta, F. A., Simonsen, L. C., Shinn, J. L., Norbury, J. W. ,1991. Transport methods and interactions for space radiation, Technical Report RP-1257, NASA.
[13] Beer, J., McCracken, K., Steiger, R. von, 2012. Cosmogenic Radionuclides: Theory and Applications in the Terrestrial and Space Environments, Physics of Earth and Space Environments. Springer-Verlag, Berlin Heidelberg.

[14] Finlay, C.C., Olsen, N. and Toeffner-Clausen, L., (2015) DTU candidate field models for IGRF-12 and the CHAOS-5 geomagnetic field model. Earth Planets and Space, 67, 114, doi:10.1186/s40623-015-0274-3.

[15] Jackman, C.H., Marsh, D.R., Kinnison, D.E., Mertens, C.J., Fleming, E.L., 2016. Atmospheric changes caused by galactic cosmic rays over the period 1960–2010. Atmospheric Chemistry and Physics 16, 5853–5866. https://doi.org/10.5194/acp-16-5853-2016

[16] Semeniuk, K., Fomichev, V.I., McConnell, J.C., Fu, C., Melo, S.M.L., Usoskin, I.G., 2011. Middle atmosphere response to the solar cycle in irradiance and ionizing particle precipitation. Atmospheric Chemistry and Physics 11, 5045–5077. https://doi.org/10.5194/acp-11-5045-2011

[17] Leung, P.T., 1989. Bethe stopping-power theory for heavy-target atoms. Phys. Rev. A 40, 5417–5419. https://doi.org/10.1103/PhysRevA.40.5417

[18] Ziegler, J.F., 1999. The Stopping of Energetic Light Ions in Elemental Matter. J Appl Phys, 85, 1249, https://doi.org/10.1063/1.369844.

[19] Kilifarska N 2017 Hemispherical asymmetry of the lower stratospheric O3 response to galactic cosmic rays forcing ACS Earth & Space Chem. 1 80–88.

[20] Kilifarska, N. A., Bakmutov, V. G., & Melnyk, G. V. 2020. The hidden link between Earth's magnetic field and climate. Amsterdam, Elsevier. ISBN 978-0-12-819346-4.

[21] Oh, S.Y., Yi, Y., 2009. Statistical reality of globally nonsimultaneous Forbush decrease events. Journal of Geophysical Research: Space Physics 114. https://doi.org/10.1029/2009JA014190

[22] Kilifarska N.A., V.G. Bakhmutov, G.V. Melnyk, Energetic particles influence on the Southern Hemisphere ozone variability, Compt. rend. Acad. bulg. Sci., 66(11), 1613-1622, 2013.

[23] Box, G. E. P., & Jenkins, G. M., 1970. Time series analysis. San Francisco: Holden Day. ISBN: 0816210942.

[24] Bishop, C., 1995. Neural Networks for Pattern Recognition. Oxford: University Press. ISBN-13: 978-0198538646.

Journal of Physics: Conference Series **2255** (2022) 012014 doi:10.1088/1742-6596/2255/1/012014

Future trends of commercial In- Orbit Satellite Servicing, Active Debris Removal and End-Of Life services

Stella Alexandrova, PhD,

S-Cosmos Ltd

Cherni Vrach, 34, Sofia - 1407

salexandrova@s-cosmos.eu

Abstract. Space debris growth has become a major threat not only to satellites as well as to the safe operations of the International Space Station (ISS). In September 2021, there were more than 4,700 operational satellites, 36,5000 space debris objects larger than 10 cm and 1 000 000 pieces in the 1cm to 10 cm range. The total mass of all space objects in Earth Orbit is 9,600 tonnes[1]. The launches of Starlink, Amazon Kuiper and OneWeb satellite constellations will increase the threat of space debris collisions.

Satellite owners, operators, space agencies and commercial players owning the mega-constellations will need to find economically viable ways to inspect, refuel, augment, extend and manage the lifetime of their satellites. Some of the emerging trends taking place in the space industry are on- orbit satellite servicing, active debris removal services and end-of life services. With the technology demonstrations on orbit of Mission Extension Vehicles (MEV-1), MEV-2 and ELSA-d missions promising In Orbit Satellite (IOS), markets are emerging and expected to reach up to $ 6.2 bln by 2030[2]. The objective of this paper is to identify the role of Bulgarian organisations in the future emerging of in- orbital satellite (IOS) and space situational services markets. Bulgaria's historical space competencies can contribute to the development of user-driven services/ solutions for space debris inspection, collision avoidance, space robotics and in- orbit servicing and assembly. By encouraging the creation of a new space eco-system in the domain of space debris and space situational awareness in South-Eastern Europe, Bulgarian companies and research organisations are bound to play an important role in the future space robotics markets.

1. Introduction

Space debris growth has become a major threat not only to satellites but also to the safe operations of the International Space Station (ISS). In September 2021, there were more than 4,700 operational satellites, 36,5000 space debris objects larger than 10 cm and 1 000 000 pieces in the 1cm to 10 cm range. The total mass all space objects in Earth Orbit is 9,600 tonnes [1]. In recent years satellite owners have launched an increased number of telecommunication satellites.

The present space debris markets are in a nascent stage of market development. Nevertheless, the stakeholders in this market are quire diverse. There are already a number of institutional and commercial stakeholders that have emerged both in the demand and supply sides

The institutional and commercial customers will be buying different services, for example, space agencies are likely to focus on buying services for clearing old satellites or parts of rockets in crowded LEO orbits and funding technology innovations . Commercial satellite operators on the other hand, will be looking at buying affordable active space robotic services for extending the lifetime of their satellites.

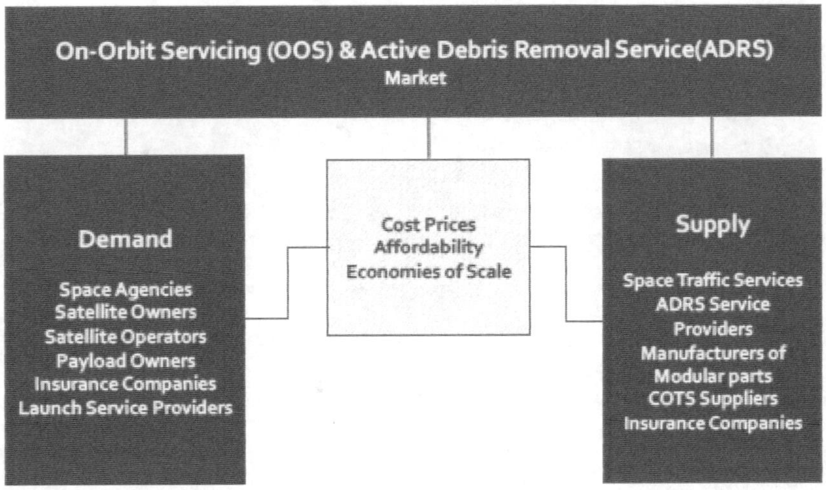

Figure 1 Stakeholders [3]

On the *demand side* companies planning to offer ADR services will have to decide whether it is worth spending on the R&D investment to develop the necessary technologies, determine their break-even point and decide on the number of satellites [4] which they will need to service per year in order to survive. One of the *biggest challenges* for ADRS/OOS companies will be to convince satellite owners and operators in the *value proposition* of the in- orbital satellite servicing and active debris services removal.

In the short term the market demand for ADRS will be driven by **institutional customers** that will invest in technology demonstration solutions which will then mature and be commercialised in the medium and long term for purely commercial customers[3]

Bulgarian companies could play a role as regards *the supply side* by offering affordable services and the ADR service providers. They could become suppliers in the provision of space traffic services, COTS suppliers and manufacturers of modular parts for satellites. The future role of Bulgarian space organisations will depend on a few factors:

- Commercial space robotic markets evolution, growth and increased demand for cost-effective and affordable space robotics services
- The emergence of a Newspace eco-system of companies in the space robotics and space situational awareness domains, Bulgaria and South- Eastern Europe will be centred in Sofia

The Bulgarian space industry will benefit from its IT and microelectronics industry. In combination with national low taxes and favourable economic conditions for attracting foreign investors, Bulgaria stands a good chance to attract foreign companies in the domain of space robotics and space situational awareness (SSA) domains and become part of the Newspace industry in Bulgaria.

2. Trends

There are a number of strategic trends taking place for satellite owners in low earth orbits (LEO) and geostationary orbits (GEO).

2.1. Low Earth Orbit trends

Satellite operators are currently exposed to falling prices for satellite capacity, aging satellites and on-going advancements in the non-GEO mega constellations. There are several reasons for increased space debris:

- Uncontrolled launches, lack of global regulation and space traffic managements services for the launched satellites since 1957
- Satellite owners are to address the space debris issue and re-assess the impact of the Kessler effect upon future space exploration
- Uncontrolled destruction of satellites in Low Earth Orbit (LEO) by the antimissile systems of countries like China, India and Russia
- Planned launches of Starlink, Amazon Kuiper and OneWeb satellite constellations have increased the threat of space debris collisions in Low Earth Orbit (LEO).

Companies, such as Astroscale successfully demonstrated the Active Debris removal (ADR) services and their commercial potential for LEO. This key innovation was noticed by OneWeb satellite owner in May 2021 and an agreement was signed between OneWeb and Astroscale.

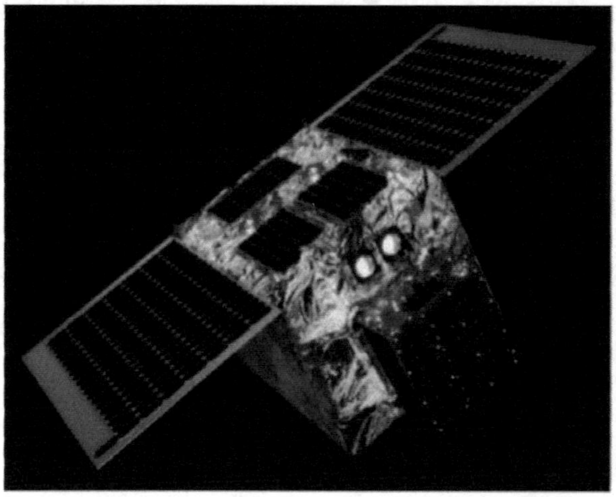

Figure 2 ELSA-d mission: showing the deployed Chaser (Servicer) with the attached Target (Client) satellite (Courtesy: Astroscale)

Clearly in LEO there will be a potential future need *for inspection, maintenance, recycling or debris removal services.*

However, a few major questions arise related to space debris in LEO : *Who will be the stakeholders willing to fund active debris removal services or end-of life ones in LEO? Are there any affordable in-orbit satellite services without governmental funding and how do they fit in the customers' business case? Who will cover the cost for space traffic services?"*. The cost for space traffic management services will have to be shared between commercial and institutional stakeholders.

Bulgarian space organisations will be able to offer space robotics services, analysis of space based data for inspection services/ development of collision avoidance prediction models and solutions using AI/ML algorithms for predicting debris collisions and protecting the national airspace.

2.2. Geostationary orbits

GEO Satellite operators will be looking at finding economically viable ways to service and extend the lifetime of their satellites at affordable prices. There are a few examples of successful in- orbit satellite servicing and active debris removal services missions, such as :

- ***MEV - 1*** (Intelsat-901) and ***MEV- 2*** **(**Intelsat 10-02**)** are both successful In Orbit Servicing (IOS) missions

Clearly the above *space missions* have demonstrated the emergence of commercial in-orbit satellite servicing market for inspection, refuelling, orbit raising and extending the lifetime of GEO satellites by at least 5 years and active debris removal services as the ELSA-d mission.
MEV-1 and MEV-2 servicing missions have been *pioneers* in the commercial In- Orbit Satellite servicing market. MEV-1 and MEV-2 missions brought *cost advantages* to satellite operators and also helped them keep customers without customer service disruption.

Figure 3: MEV-2 in orbit (Courtesy: Northrop Grumman)

The success of these missions *created* an opportunity for the emergence of *new in orbit and debris removal servicing* markets in which the USA is a leader. However, there is ***a need*** for affordable inspection, maintenance, servicing, refuelling and recycling of old satellites and IOS, ADR and EOL services in Europe.

3. Bulgarian space research and technology innovation hub

A ***space research and technology hub*** could be created in Bulgaria. It could attract space robotics and IT companies offering a variety of space robotic and SSA services and solutions, related to automation, space based data processing, AL/ML learning, computer vision, modular satellites and payloads, wireless power relevant for developing IOS, ADR and EOL services.

For Bulgaria this is a ***new opportunity*** to create a ***"public-private-partnership"*** with the Bulgarian government, space research institutes and companies. Creating a PPP following the model of ESATIA Advanced research in Telecommunication Systems (ARTES) program[5] could be quite attractive.

With minimal public investment, the Bulgarian space industry will have the opportunity to develop new products and services for upstream and downstream space applications and mature their technology solutions.

Through cooperating under PPP space institutes and companies can receive funding for achieving higher Technology Readiness Level (TRL) of their products and solutions. At the same time they can commercialise their products and services, test their products with new customers and enter the IOS, SSA, ADR and EOL markets. The benefits will be the following:

- High risk investments in disruptive technology products and solution can be funded by public money in cooperation with Bulgarian space institutes and companies
- Bulgarian space research institutes and companies, will have the opportunity to create pilot projects with commercial customers and mature the TRL levels of their products and services
- "end-user" requirements and key performance indicators (KPI) for the proposed space products and solutions could be easily integrated in the pilot projects

Commercialisation of the space robotic and SSA products and services will be possible, as the commercial customer business cases would have already tested them through successful pilots. In addition, direct economic benefits, such as disruptive technology innovation, development of new space applications and increased employment in the Bulgarian space industry will be beneficial for the Bulgarian economy.

4. Space robotic markets

Space robotics markets are in a nascent stage of evolution. Therefore there are a number of uncertainties for figuring out cost-effective ways regarding the design, manufacturing, assembly and servicing of satellites both in LEO and GEO orbits. For example, space agencies will be likely to focus on buying services for clearing old satellites or parts of rockets in crowded LEO orbits[6]
Both commercial and institutional stakeholders will face the challenge of identifying the actual market demand for in-orbit satellite servicing, active debris removal and end-of life services. Certain studies, such as the NSR one [2] show that In-Orbit Satellite services (IoS) and SSA cumulative revenues will reach \$6.2 bln by 2030. The space robotic markets can be classified into the presented sub-categories

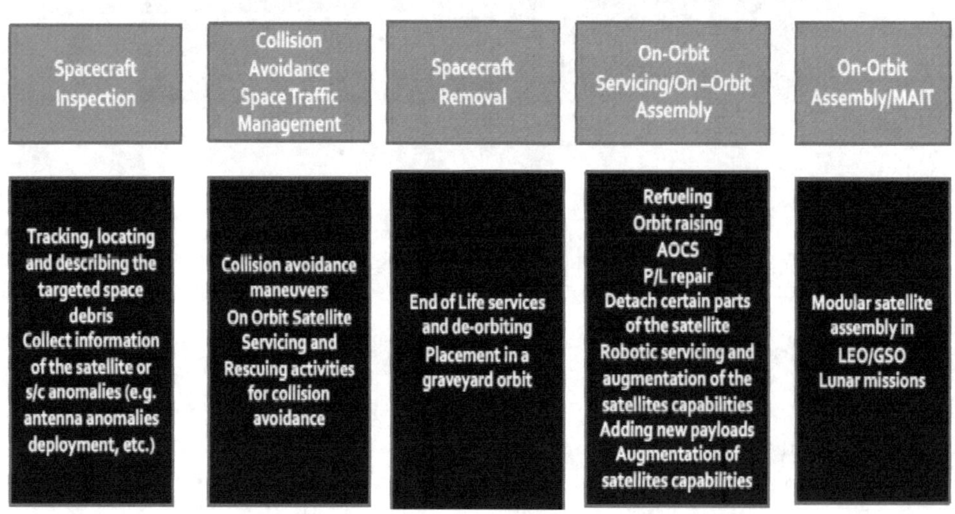

Figure 4 Space Robotic Markets [6]

The above space robotic markets are quite diverse but they all will need *cost-efficient* and *affordable* services. In addition certain institutional or commercial customers may buy urgent services like

inspection, refueling or *orbit raising.* Access to affordable services with high TRL will be positive factors as satellite owners will be able to keep their customers on board and sustain their profitability and communicate to insurance companies real-time information about the satellite status.
Nevertheless, there will be a number of uncertainties in the *future space robotic markets*. related to the following:

- Unknown commercial market demand for IOS and ADR services
- Unknown institutional market demand for IOS and ADR services
- Unknown customers
- Diversity of the space markets

The above uncertainties may be overcome once the markets start to develop. Clearly there is an ***opportunity*** to develop a user-driven services and solutions for space robotic removal services for LEO/MEO/GEO satellites.
Bulgarian research and industrial organizations shouldn't miss this opportunity and prepare in advance to offer affordable and on- time services in the space robotic domain.

4.1. Future technology trends

Clearly in LEO there will be a potential future need *for inspection, maintenance, recycling or debris removal services.* Companies like Astroscale are developing the ELSA-M servicer that will remove a number of satellites only with one servicers as presented in the figure below.

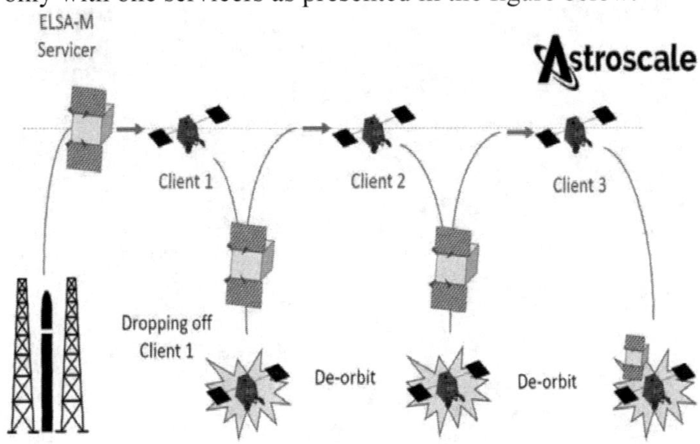

Figure 5 ELSA-M (Courtesy: Astroscale)

Others are already offering satellite inspection services, such as the SCOUT company that is developing the SCOUT OVER-sats for rendezvous & proximity operations (RPO) and conjunction assessments.

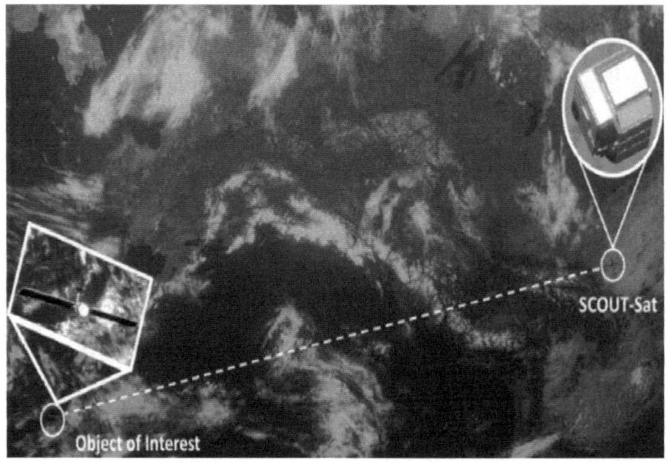

Figure 6 SCOUT-OVER sats (Courtesy: SCOUT)

In addition, certain European companies, such as iBOSS GmbH are developing concepts for modular satellites. Future concepts suggest modular satellite systems, for example, satellites based on modular design allowing for efficient and most flexible assembly and purpose changes on ground, assembly or servicing in orbit, etc., as well as new concepts of operations and innovative business models.

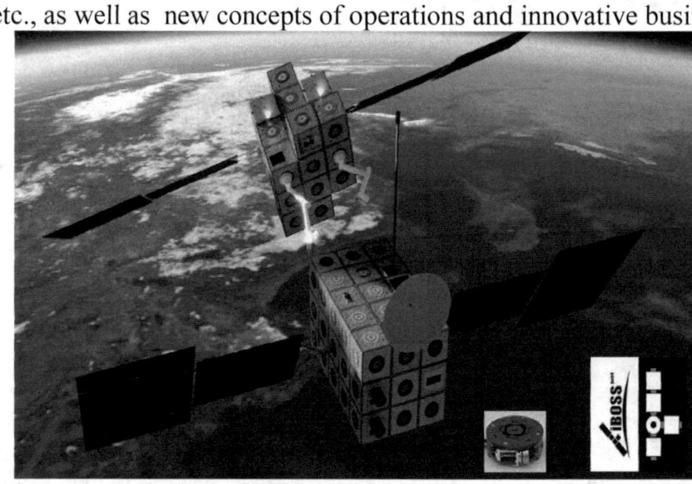

Figure 7 Modular satellite concepts (Courtesy: iBOSS GmbH)

The idea is to compose satellites systems of *a variety of pre-qualified and pre-tested standardized building blocks*. These systems would then be supported by a logistics infrastructure comprising, warehouses, modules and building blocks and assembly and servicing infrastructure elements assisted by robotics [6]

Multi-servicing satellite removal servicers, modular satellites and real-time conjunction assessment are some of the current technology trends taking place in the space industry.

5. Conclusions

MEV-1, MEV-2 and ELSA-d missions technology demonstrations have opened new In Orbit Satellite (IOS) and Space Situational Awareness (SSA) markets and expect to reach up to $ 6.2bln by 2030. European companies haven't yet performed technology demonstration mission of in-orbit satellite servicing neither in LEO nor GEO orbits.

Journal of Physics: Conference Series **2255** (2022) 012014 doi:10.1088/1742-6596/2255/1/012014

There is a niche and an ***opportunity*** to develop a user-driven service and solutions for space robotic removal services for LEO/MEO/GEO satellites. *Bulgarian research and industrial organizations* shouldn't miss this opportunity and ought to be ready to offer affordable and timely services in the space robotic domain.

Bulgarian companies could play a role on *the supply side* and contribute to offering affordable services and the ADR service providers. They could become active suppliers in the provision of space traffic services, COTS suppliers and manufacturers of modular parts for satellites. Bulgaria's space competencies could contribute to the development of user-driven services/ solutions for space robotic inspection, collision avoidance, space robotics and in orbit servicing and assembly.

The Bulgarian space industry will benefit from its IT and microelectronics industry. In combination with national low taxes and favourable economic conditions for attracting foreign investors, that will draw foreign companies to become part of the Newspace industry in Bulgaria in the domain of space robotics and the space situational awareness domains. Thus, in South-Eastern Europe will emerge a Newspace eco-system of companies in space robotics and space situational awareness based in Sofia.

A space research and technology hub could be created In Bulgaria for attracting space robotics and IT companies for offering a variety of space robotic and SSA services and solutions that will be related to automation, space based data processing, AL/ML learning, computer vision, modular satellites, payloads, wireless power relevant for developing IOS and ADR services, developing SWE prediction models and protecting the national airspace. For Bulgaria this is an opportunity to create a ***"public-private-partnership"*** with the Bulgarian government, research institutes and space companies. By creating a PPP with pilot projects following the model of ESA ARTES program, higher TRL of products and solutions and successful commercialisation of space robotic and SSA services will be possible.

By encouraging the creation of a new space eco-system in the domain of space robotics and space situational awareness in South-Eastern Europe, Bulgarian companies and research organisations will begin to play an important role in the future space robotic markets.

References

[1] ESA, *Space Debris by the numbers* https://www.esa.int/Safety_Security/Space_Debris/Space_debris_by_the_numbers

[2] NSR, In orbit servicing &space situational awareness markets, 4th edition, https://www.nsr.com/?research=in-orbit-servicing-space-situational-awareness-markets-4th-edition

[3] S.Tkatchova. (2020, August). *The commercial future of orbital services . Ruimtevaart* , pp. 17-19.

[4] S.Tkatchova (2018). *Emerging Space Markets*,Ch6. *Space Debris Mitgation,* pp.99-105

[5] ESA, About ARTES, https://www.esa.int/Applications/Telecommunications_Integrated_Applications/ARTES/About_ARTES

[6] Alexandrova, S., J.Kreisel, T.Schervan, *Measuring the Direct and Indirect Economic Benefits of Commercial Active Space Debris Removal Services (ADRS) based on the utilisation of disruptive future modular satellite systems* 71st International Astronautical Congress (IAC)- The Cyber Space Edition, IAC-20, E6,2,10 x ID58656,

Journal of Physics: Conference Series **2255** (2022) 012015 doi:10.1088/1742-6596/2255/1/012015

Analysis of hyperspectral and multispectral reflectance spectra in the Black Sea coastal area near the Danube delta: comparison of PRISMA and Sentinel-2 observations

Irina Gancheva

Sofia University St. Kliment Ohridski; 5 James Bourchier blvd. 1164 Sofia, Bulgaria

E-mail: irina.gancheva@phys.uni-sofia.bg

Abstract. In this study we investigate the possibility to distinguish between different water types in the Danube delta and the Black Sea coastal zone based on the reflectance spectra. For this we use hyperspectral satellite images from PRISMA (PRecursore IperSpettrale della Missione Applicativa) and multispectral images from Sentinel-2 MSI (MultiSpectral Instrument) in July 2020.
The visual inspection of the available acquisitions differentiates between four aquatic types: lagoons; river and stream plumes mixing with marine waters; turbid and optically complex coastal waters; and optically clean waters away from the shore. For each of the four types we determine the characteristic averaged reflectance spectra from PRISMA and Sentinel-2 for the spectral range from 400 nm to 2500 nm. It is found that Sentinel-2 spectra are very similar for the 4 water types, in contrast to the PRISMA spectra which give substantial additional information.
Further we analyse the gradual modification of the characteristic hyper- and multispectral spectra from the shoreline to the open sea (~25 km distance) in 15 locations passing through areas of varying turbidity. We demonstrate that the intensity of surface reflectance from the hyperspectral instrument decreases gradually with distance from shoreline, clearly showing the transition zone between riverine and marine water. The multispectral reflectance spectra for the same study areas did not present such consistent behaviour.
The presented results demonstrate the benefits of hyperspectral over multispectral images for turbid aquatic regions in the Black Sea coastal zone. They show that with little requirements regarding pre-processing and computational resources hyperspectral data can contribute greatly to classification of water types, in respect of their turbidity.

1. Introduction

The optical remote sensing of the Danube delta entering the Black Sea is a challenging task due to the optical complexity of the aquatic system: on one side there are highly turbid riverine waters mixing in Case-2 marine waters [1] and on other, there are many lagoons and inland brackish lakes. The region is protected area, requiring timely and reliable monitoring of the water quality, which highlights the importance of advancing the available sampling methods and testing new data sources [2]. In such hard-to-reach areas, where the in-situ sampling is often sparse, spaceborne datasets provide much needed observations for water quality assessment.

In the last 50 years multispectral optical missions have advanced significantly in terms of temporal and spatial resolutions [3,4] and their dataset have contributed a lot to understanding world oceans and inland water body processes. In recent years, data from the Sentinel missions have shown as very important in that regard [5].

The immense potential of hyperspectral remotely sensed data has been recognized by the scientific community and research topics cover wide range, from agriculture and land cover mapping to water quality monitoring and coastal zone management [6]. Hyperspectral sensors provide large number of bands with good spectral resolution of 6.5 to 25 nm and wide wavelength range from shortwave Ultraviolet (UV) and Visible (VIS), up to Short Wave InfraRed (SWIR) [6], covering many application domains [6,7].

Hyperspectral images have proven as very beneficial in particular for water quality assessment, but an important constrain for the operational monitoring remains the poor availability of such missions due to the high cost of hyperspectral sensors and the significant amount of data volume leading to the need of high computation resources. In this case multispectral missions provide greater efficiency with short revisit time and good spatial resolution. Nevertheless, hyperspectral imaging contributes greatly to increasing the estimation accuracy of important water quality variables (such as chlorophyll-a, Total Suspended Matter (TSM) compared to multispectral instruments [7]. Also, hyperspectral data has proven important in the indication of new variables of interest (especially in the context of characterization of particulate and dissolved matter: type, size of suspended particles; types of pigments; but also in identification of inorganic pollutants; cyanobacteria, etc.) [7].

There are very few studies of the Black Sea basin, analyzing hyperspectral datasets. Lavrova et al. [8] did joint investigation of different datasets, highlighting the benefits of hyperspectral images with study area covering part of the northeast Black Sea and Azov Sea. The study looked at the optical properties of moderately turbid and productive coastal zone waters and inland water bodies, comparing the reflectance spectra. The hyperspectral data has proven as very important for detecting and discriminating different types of anthropogenic and biogenic pollution in coastal zones. Ghezehegn et al. [9] used handheld hyperspectral radiometer for determining Chlorophyll-a concentrations and compared the results with satellite observations obtained from MODIS (Moderate Resolution Imaging Spectroradiometer), showing good correlation between both.

PRISMA is a novel mission launched in March 2019 and is developed and maintained by the Italian Space Agency ASI (Agenzia Spaziale Italiana). It is flying on Sun synchronous low Earth orbit and has nominal orbit re-visit time of 29 days with the possibility of reducing it to 7 days with off-nadir viewing for a specific target. The imaging spectrometer has two separated instruments – the Visible and Near-InfraRed (VNIR) spectrometer (400 nm to 1010 nm) with 66 bands and bandwidth lower than 11 nm and the SWIR detector (920 nm to 2500 nm) with 174 bands and bandwidth lower than 15 nm. The single image size is 30×30 km and the pixel size is 30×30 m for the VNIR and SWIR channels and 5 m for the PAN-chromatic channel [10]. The PRISMA acquisitions are available to the scientific community since May 2020 freely after registration on https://prisma.asi.it with a catalogue of archived acquisitions and the possibility of requesting acquisitions for a specific area of interest.

There are already some very promising results based on PRISMA and Sentinel-2 analysis for various water regions.

Giardino et al. [11] presents a first assessment of radiometrically calibrated Level 1 Top-of-Atmosphere (TOA) radiance in visible and near-infrared from PRISMA. The datasets are compared with the TOA radiances simulated with a radiative transfer code, starting from in situ measurements of water reflectance. Additionally, the PRISMA data is compared as well to TOA radiance from concurrent Sentinel-2 Level 1C MSI images. The study points were distributed over different aquatic regions, coastal and inland turbid water basins. Results show that PRISMA compared with both, in-situ simulated datasets and Sentinel-2 L1C data, show TOA radiance of same magnitude, concluding that synergetic use is beneficial.

Niroumand-Jadidi et al [12] compared Sentinel-2 and PRISMA Level 2D data for inland water applications, choosing a turbid lake region in Italy (Lake Trasimeno). From both products they compared the remote sensing reflectance and retrieved water quality parameters (Total Suspended Matter (TSM), Chlorophyll-a (Chl-a), and Colored Dissolved Organic Matter (CDOM)) and performed consistency analysis of the prepared datasets. The cross-sensor comparison of the remote sensing reflectance (Rrs) data showed strong agreement at different wavelengths. The consistency analysis showed good agreement between Sentinel-2 and PRISMA derived water parameters.

In the context of newly emerging satellite missions and the need of timely developed applications for water quality monitoring, having the possibility for quick assessment of a region of interest has great value. Here we demonstrate that hyperspectral data has some considerable advantages over multispectral images for initial analysis and categorization of marine coastal regions, impacted by fluvial discharge.

The domain of interest is the entering of the Danube in the Black Sea through three main branches. The large river discharge here makes the marine area unique: the mixing between water of marine and riverine origin forms a transition zone. In this region it is very important to understand the movement of the frontal zone which is very dynamic and depends on the local meteorological conditions. In such cases the monitoring from space would prove beneficial in order to identify the riverine water extend.

2. Methods and study domain

In this study, we analyze reflectance spectra of Sentinel-2 MSI and hyperspectral PRISMA images in the spectral range from 400 to 2500 nm. The region of interest covers the northwest area of the Black Sea coastal zone including the Danube delta. Due to the limited availability of PRISMA images it is difficult to find overlapping snapshots of Sentinel 2 and Prisma. Fortunately, there are five available PRISMA cloudless acquisitions for this region from the 4th July 2020. The tiles are acquired successively in descending orbit direction at 9:06 am. The nearest in time Sentinel-2 tiles cover the area of the PRISMA acquisitions in 6th July 2020. The tiles are acquired successively in ascending orbit direction at 8:56 am.

The PRISMA products are Level 2D, they are atmospherically corrected, geo referenced and are available for download in he5 format. The Sentinel-2 products are Level 2A, atmospherically corrected and in cartographic geometry (UTM/WGS84 projection). Corresponding to the growing need of the scientific community, new tools are developed for managing and analysis of hyperspectral data. EnMAP-Box (https://www.enmap.org/) is a user-friendly QGIS plugin, freely available for download. It can manage both PRISMA and Sentinel-2 data, as well as images from some other common multi- and hyperspectral missions.

Figure 1 shows true color Red-Green-Blue (RGB) composites from both sensors with Sentinel-2 composite at R: 665 nm, G: 560 nm, B: 492 nm and PRISMA composite at R: 660.2894 nm, B: 562.759 nm, G: 492.7201 nm. The channels in the two RGB composites are chosen to be as close as possible having in mind the sensor construction.

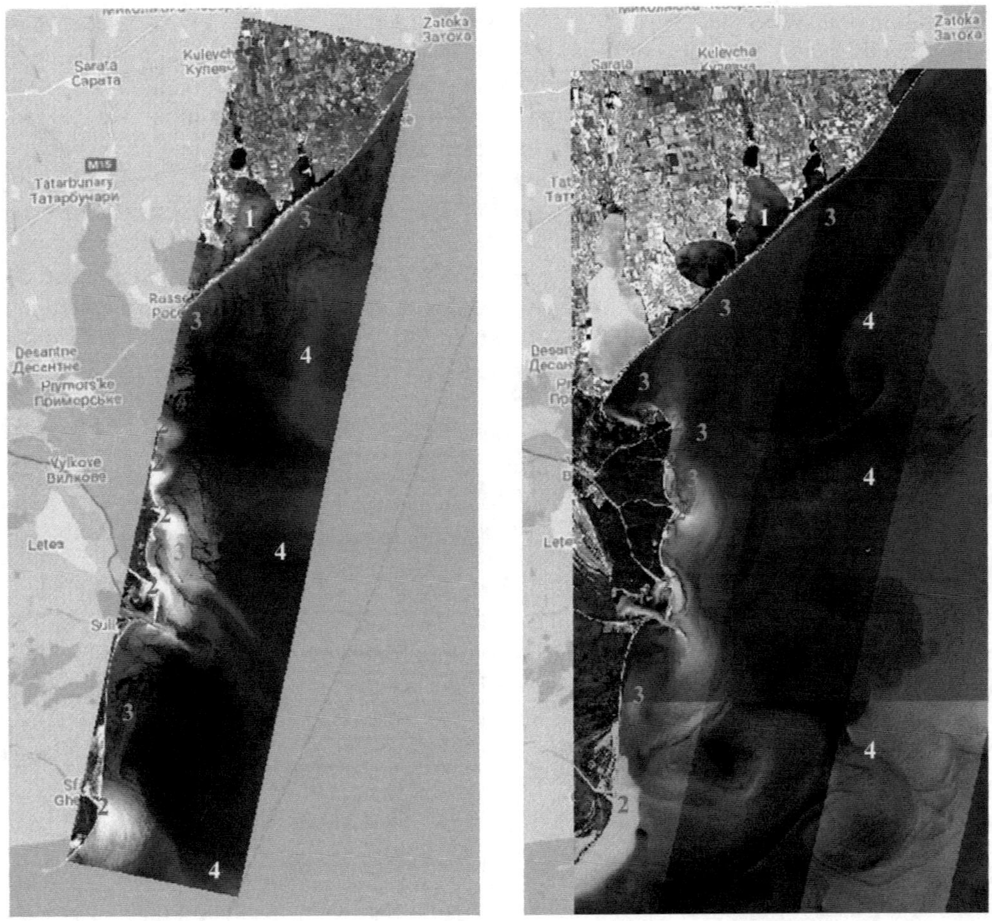

Figure 1. RGB composites of PRISMA (left) and Sentinel-2 (right) of the entire study region. The numbers indicate different types of water: 1 – lagoons, 2 – river and stream mouths, 3 – turbid coastal waters, 4 – clean marine waters.

In both images we could visually identify four types of water: lagoons, river and stream mouths, turbid coastal waters, and clean waters at max. 28 km distance from the shoreline.

The first type is relatively turbid marine-coastal waters, covering parts of the Shahany, Alibei and Budaki Lagoons in the Tuzlovski Lagoons National Park, in the southern Bessarabia, Ukraine (marked with white number 1 on figure 1). These are shallow brackish water bodies with average depth 1-1.5 m separated from the Black Sea through sandbars. The lagoons play an important role in maintaining biodiversity in the Black Sea coastal ecosystems. It is a breeding site for many bird species, some of them also endangered and hosts some notable fauna, including species from the Red Data Book of Ukraine [13].

The second type of water is riverine: the sources are six rivers and stream mouths, including the Sfantu Gheorghe Branch, Sulina Branch, Novostambul's'ke and Ochakivs'ke River mouths (red number 2 on figure 1). The inflowing river plumes are well mixed and thus appear with higher sediment concentrations. They are similar to the lagoon type regarding their reduced salinity, but the difference is in the movement – lagoon water is stagnant.

The coastal type includes optically complex and turbid water with increased sediment concentration (pink number 3 on figure 1).

The clean water type could be seen at approx. 8 km to 28 km distance from the shoreline and is characterized by low turbidity and optically clean blue waters (yellow number 4 on figure 1). The characteristic spectra for these 4 types of water are determined for both instruments – MSI and PRISMA. They are obtained by averaging the spectra of a significant number of pixels in each region. The number of pixels for averaging is chosen depending on the size of the area it should represent and pixels giving outlier values are excluded from the averaging.

Our idea is to find how the water type changes from the shore to the open sea. In the next step we compare the reflectance spectra measured by Sentinel-2 and PRISMA for 15 points located at varying distance from the shore, starting from the Novostambul's'ke River mouth, reaching to optically clean waters 25 km away from the shoreline. This is illustrated in figure 2. Each presented spectrum gives the average spectrum of 9 pixels around each point on figure 2. The points are chosen in a way that all 9 pixels are cloudless. During the selection small changes in the location of approx. 15 pixels were done to ensure that the chosen area is representative.

Investigating the spectra in these locations, we aim to identify with more precise the different water types of the region, visually tracking the reflectance change over distance. Moreover, we demonstrate that with little pre-processing of available hyperspectral data and calculation of secondary water quality parameters, one can extract information about the characteristics of water in a region of interest.

One of the main concerns is the fact that both images were not acquired in the same moment but rather with 48 hours time-lag. The visual inspection showed that both images reveal similar surface structures with comparable spatial extend. Additionally, the points for analysis were chosen in a way to correspond to very similar optical surface conditions, so that both spectra are comparable.

We evaluated the meteorological conditions for the time-period of both acquisitions. Meteorological data is taken from the European Center for Medium-Range Weather Forecast (ECMWF) ERA5 reanalysis [14]. The total precipitation for the entire Danube Delta region was analyzed for a time frame of 4 days before the PRISMA acquisition and three days after. No significant rain events were observed, leading to the conclusion that the plume volume on both days does not vary significantly.

The wind speed and direction are visualized on the second line of figure 2. The arrows give the wind direction, and the wind speed is presented on the grey line in [m/s]. The timing of each the acquisitions is marked with a black vertical line.

Four days before the PRISMA overpass the wind speed is low with variable direction. At the time of the overpass the wind speed drops less than 1 m/s and in the next two days it increases gradually to maximum of 6.5 m/s from N / NW. During the Sentinel-2 overpass the windspeed is approximately 5 m/s from NE which corresponds to a moderate breeze.

We conclude that the comparison of the two observations is reasonable despite the slight difference in the wind conditions. This is proven by the visual inspection which shows that both images reveal similar surface patterns.

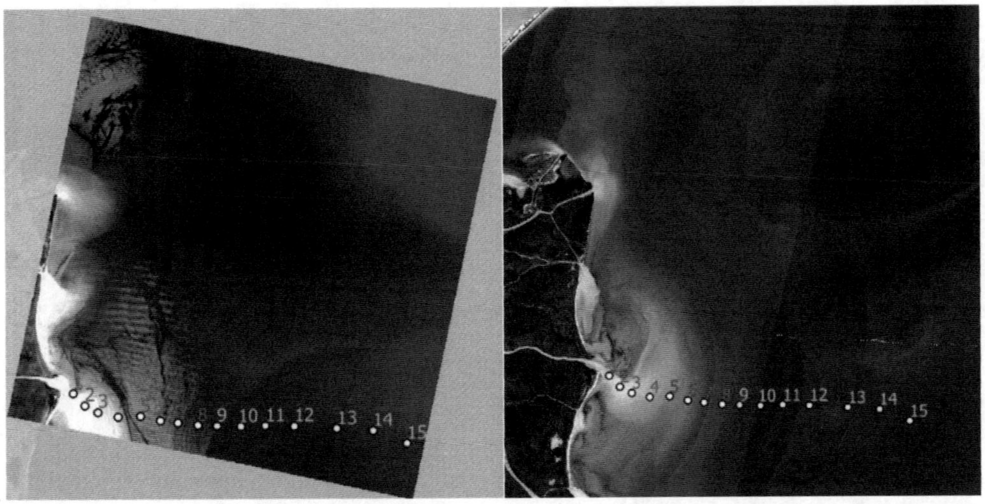

a

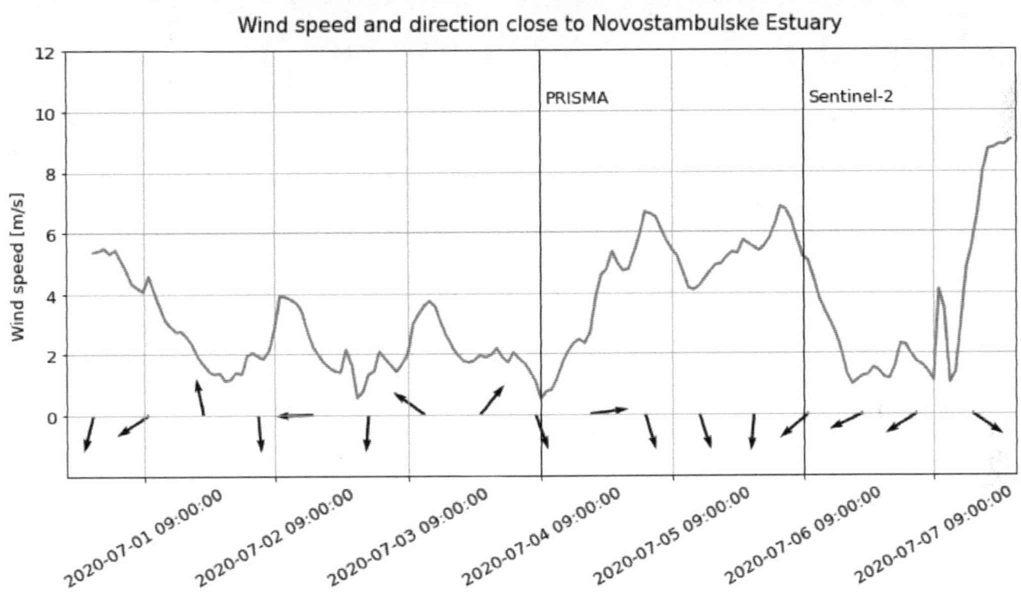

b

Figure 2 a: RGB composites of PRISMA (left) and Sentinel-2 (right) with the 15 study points for which the profiles are given. Points 1-8 represent riverine and turbid coastal waters and points 9-15 represent clear marine waters.
b: Hourly data for wind speed and direction at the Novostambulske River estuary from 1.7.2020 until 7.7.2020. The acquisition times of PRISMA and Sentinel-2 are marked with vertical black lines.

3. Results and discussion

3.1. Characteristic spectra.

Figure 3 gives the characteristic Sentinel-2 (blue) and PRISMA (yellow) spectra of each of the four predefined regions: lagoons, river and stream mouths, turbid coastal waters and clean blue waters.

Journal of Physics: Conference Series **2255** (2022) 012015 doi:10.1088/1742-6596/2255/1/012015

a: Lagoons

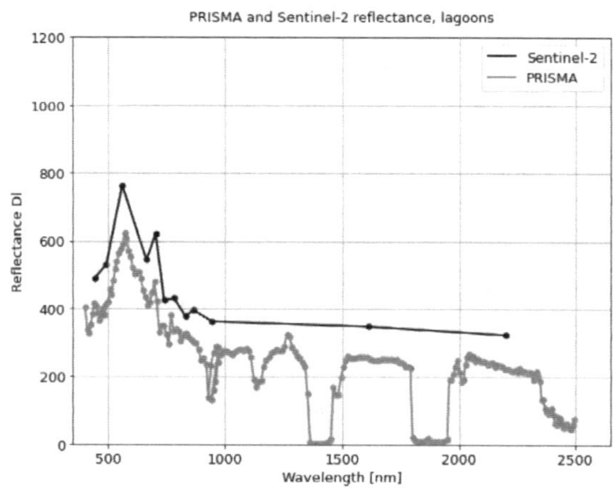

b: River and stream mouths

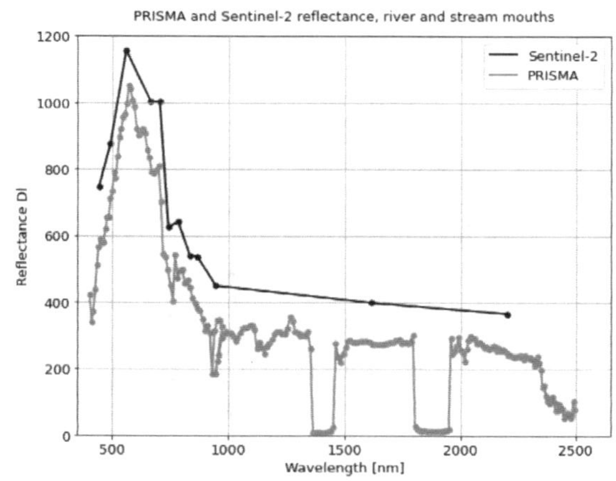

c: Coastal waters

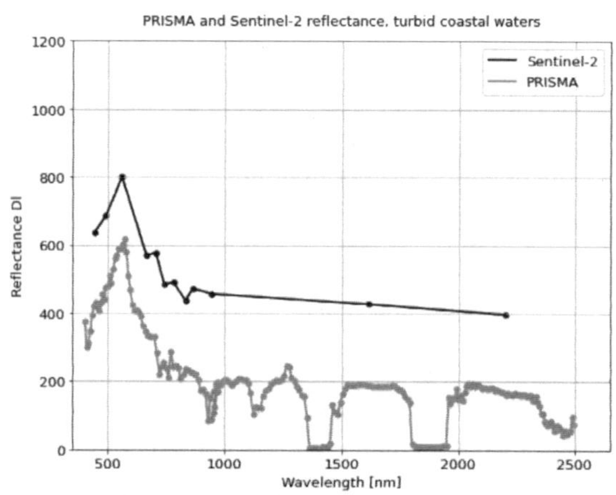

d: Clean waters

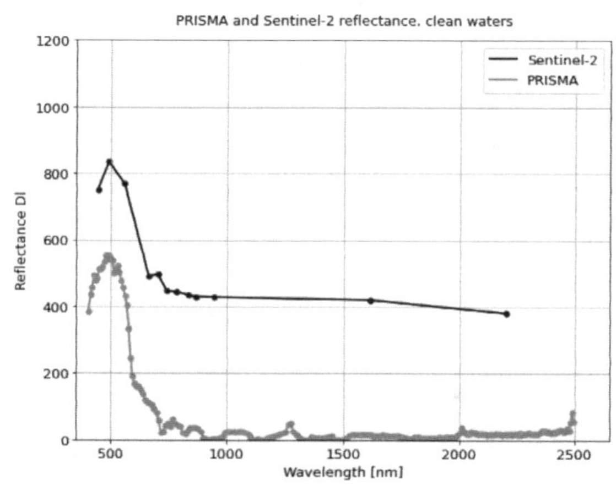

Figure 3. Local reflectance spectra from Sentinel-2 (blue line) and PRISMA (yellow line) for the four predefined areas of interest – **a:** Lagoons, **b:** river mouths, **c:** coastal turbid waters, **d:** clean waters.

The four graphs in the figure 3 reveal that for all water types the Sentinel-2 spectra have higher reflectance values compared to PRISMA. However both spectra peak in the same wavelength region (~600 nm). The deviation between the two spectra can be partly explained with the different atmospheric correction applied on the products.

At shorter wavelengths the spectrum is similar but at longer waves in the range where Sentinel-2 does not measure, the Prisma spectra gives substantial additional information. Such are the local maxima at 1260-1270 nm and gradual decrease followed by an increase in the reflectance after 2300 nm, visible in all four characteristic types (in region d: clean waters, the reflectance in SWIR drops to very low values after the local maxima mentioned above; the increase in the last 50nm of the spectrum is visible). Very clear are seen the water vapour absorption bands between approx. 1360 – 1450 nm (the cirrus absorption bands) and 1810 – 1950 nm, where the reflectance intensity in all figures drops to zero and the local minima at 930-950 nm and 1120-1150 nm are visible.

In the visible range the PRISMA spectra reveal global maximum at 571 nm and Sentinel-2 at the green band centred at 560nm for the lagoons, river mouths and coastal water regions. For the clean water region, it is at 500 nm for PRISMA and 492 nm (blue band) for Sentinel-2. Having the precise reflectance distribution, especially in the visible spectrum is important for the calculation of secondary water quality variables, such as chlorophyll-a, CDOM and turbidity concentrations. Recent studies comparing PRISMA and Sentinel-2 data (water quality products [12] and TOA reflectance with in-situ data [11]) for inland turbid waters have demonstrated good agreement in the values derived from both data sets, leading to the conclusion that synergic use of both sensors for aquatic applications is beneficial [11].

Comparing the value ranges of the PRISMA spectra for all four water types clearly distinguishes the different water: the highest surface reflectance is observed for the type in the river mouths with maximum values in the visible and near-infrared spectra. In the short-wave infrared wavelengths the lagoons and river mouths have comparable spectra. The turbid coastal water type shows similar distribution, but with lower reflectance values. The increased reflectance in river mouths can be explained with the high concentration of optically active particles and sediments, which are carried by the river stream into the marine environment. The dropping reflectance in VNIR and SWIR for the clean water region corresponds to previous marine and inland water measurements [15], which demonstrate strong absorption of water in the long wavelength region. For water, characterised by increased turbidity and concentrations of optically active components, the VNIR and SWIR spectra are more informative for the determination of the water type. In aquatic applications SWIR can be used for determining the spectral response of blue-green algae blooms and turbid regions.

Regarding the Sentinel-2 spectra, it is seen that the four different types do not differ as much as the PRISMA ones. The Sentinel-2 spectra are informative for the particularities of the four different types, but only in the visible wavelength range up to band 8A, which gives the narrow near-infrared. Up to this band the curves show the same general propagation, which is also indicated by the PRISMA spectra. In the SWIR region they mismatch the scale and values distribution of the PRISMA spectra. The water type and turbidity appear not to have influence on the Sentinel-2 reflectance spectra in the long wavelengths and no surface signatures can be distinguished. In this context multispectral data from Sentinel-2 is important for the calculation of secondary water quality parameters such as chlorophyll, turbidity and CDOM, however it is difficult to base analysis solely on the reflectance spectra.

3.2. Profile spectra.

The satellite observations have limitations near the coast where the image could be influenced by the reflectance from the bottom or other local conditions. In this chapter we investigate the possibilities to identify different water types from the shoreline to the open sea and discuss the differences between hyper- and multispectral images.

Figure 4 gives the reflectance spectra in 15 areas for both instruments, as shown in figure 2a and each spectrum is represented with a different colour. For better visualisation the spectra are separated in two figures: points 1 to 8 correspond to fluvial and more turbid coastal waters plotted on figure 4a (PRISMA) and 4c (Sentinel-2); points 9 to 15 correspond to optically clean marine waters, shown on figure 4b (PRISMA) and 4d (Sentinel-2). This separation makes the analysis clearer.

First thing to be noted is that the PRISMA spectra in the 15 points are quite different form each other and clearly grouped in three characteristic patterns. In contrast the Sentinel-2 spectra do not differ substantially. This is a firm prove that the hyperspectral image gives more options to categorize the water type.

The reflectance spectra of the PRISMA instrument for the fluvial and turbid coastal regions reveal very similar distribution up to point 6 with reflectance gradually decreasing. The separation is visible best in the wavelength region from the green reflectance maximum at 550-570 nm to the water vapour absorption window at 940 nm. For that region one clearly recognises the maximum reflectance for the point positioned in the fluvial plume and the gradual decrease moving away from it. The last two points from figure 4a reveal reflectance spectra more typical for the optically clean waters with point 7 being transitional between both water types.

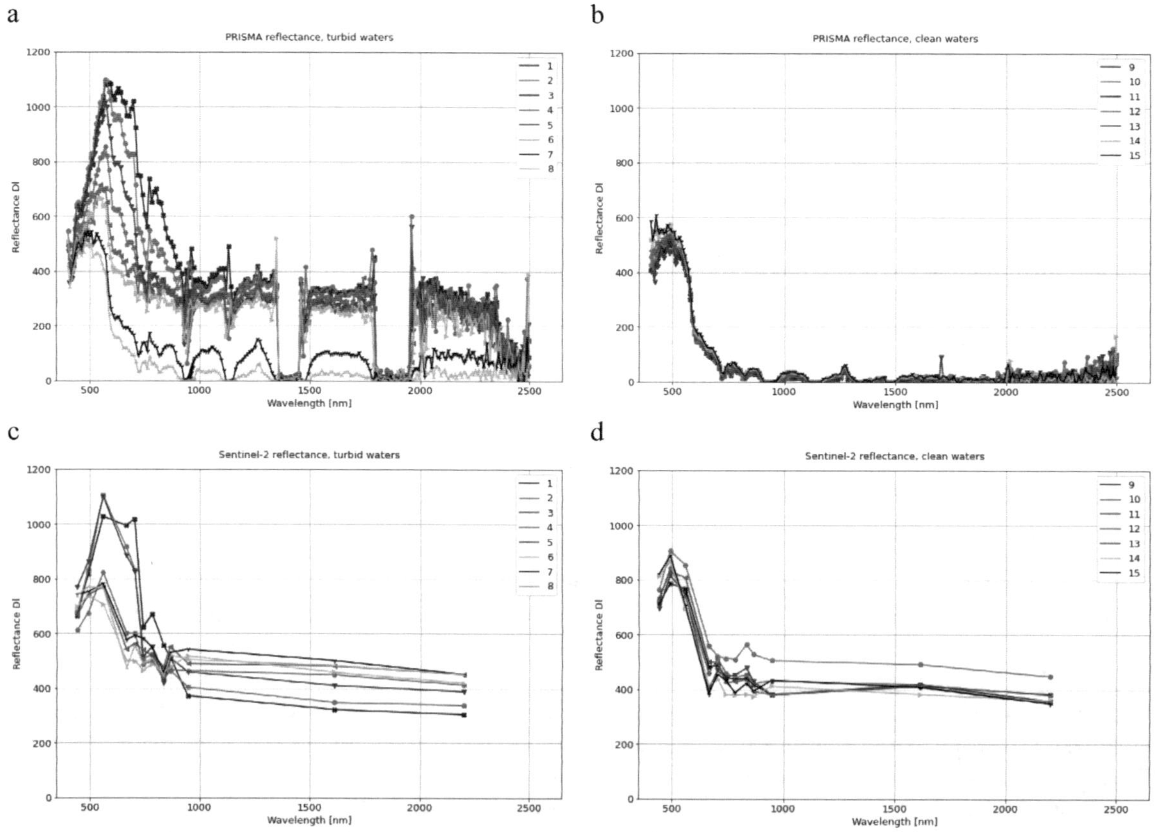

Figure 4. Reflectance spectra for the 15 points shown on figure 2a.
a: PRISMA riverine and turbid coastal waters (points 1-8); **b:** PRISMA clear marine waters (points 9-15); **c:** Sentinel-2 riverine and turbid coastal waters (points 1-8); **d:** Sentniel-2 clear marine waters (points 9-15).

Figure 4b shows the reflectance spectra from PRISMA for optically clean waters (points 9 to 15). It becomes evident that all spectra have very similar distribution with few exceptions mainly in the SWIR region. The exact analysis also shows that points closer to the coast have slightly lower reflectance compared to the ones at greater distance, however with some exceptions and mainly in the visible wavelengths range. It can be concluded that the regions of clean marine water have very similar spectra and no further specifications of the water type can be done, especially for the VNIR and SWIR wavelengths.

Figure 4c and d give the reflectance spectra of Sentinel-2 for the points of interest. They are again separated between two figures with spectra from the fluvial plume and turbid waters plotted on figure 4c. The colour coding is the same as for the PRISMA spectra. The first three graphs, corresponding to the fluvial plume region are separated by the rest by their increased reflectance. The other points have generally lower values, however no consistent relation between the distance from the shore and the intensity of the reflectance can be determined. Analogous is the case with the reflectance spectra from the points located at clean waters. Their reflectance intensity is partly higher than the fluvial and turbid points, marking no separation between the regions. These results are in contrast with the PRISMA spectra, where well distinguished separation could be seen.

The increased reflectance in the turbid and riverine regions can be referred to the presence of optically active particles with strong absorbance along the entire wavelength range from 400 to 2500 nm. Moving away from the shore and thus the reduction of the riverine effect on marine waters decreases the reflectance intensity, contributing to the possibility to distinguish between different water types – river plumes and such with varying turbidity.

4. Conclusion

In this study we compare two acquisitions of PRISMA and Sentinel-2 reflectance spectra in July 2020 over an area in the Danube delta and northeast Black Sea coastline, in attempt to identify water types with various characteristics. On the optical images by visual inspection four types are distinguished: lagoons, river and stream plumes mixing with marine waters, turbid coastal waters and optically clean waters at distance 8 – 29 km from the shoreline. The averaged reflectance spectra from both instruments for each of these regions was determined and specific characteristics are derived. The riverine plume water has the highest reflectance values, and the lagoon and coastal regions show very similar surface reflectance. The real advantage of the hyperspectral sensor over MSI is visible in VNIR and SWIR for turbid waters. While Sentinel-2 shows very similar reflectance for all four types, PRISMA gives substantial additional information.

Further we investigate how to determine the riverine water extent to the open sea and the frontal zone between fresh and saline water. We analyse the reflectance spectra in 15 locations positioned normally to the shoreline, starting from a river mouth and ending at the clean waters region. The PRISMA spectra of the points reveal gradual decrease in direct relation to the distance from the shore for the types of river plume and turbid coastal waters. Further away from the shoreline the water turbidity decreases. One can clearly identify the extent of the riverine water away from the coast. The spectra of the Sentinel-2 instrument show increased reflectance only for the first three point, located in the river plume. No further correlation could be determined for the rest of the spectra.

These results highlight the benefits of hyperspectral satellite images over multispectral data for marine and coastal regions. Multispectral spectra from Sentinel-2 fail to represent unambiguously different water types near the shore based only on reflectance spectra analysis. In contrast hyperspectral data from the PRISMA mission gives a promising result, that even with little requirements regarding pre-processing and computational resources, initial statements about the influence of riverine waters and turbidity levels in a region of interest can be done. Nevertheless, further analysis with hyperspectral acquisitions from different times of the year should be done, in order to validate the results for various observation scenarios and generalize the conclusions.

References

[1] Mobley C, Stramski D, Bissett P and Boss E 2004 Optical Modeling of Ocean Waters: Is the Case 1 - Case 2 Classification Still Useful? *Oceanography* **17** 60–7

[2] Institute (NGI) N G Razelm-Sinoe Black sea Lagoon in Danube Delta Biosphere Reserve *Nor. Geotech. Inst. NGI*

[3] Blondeau-Patissier D, Gower J F R, Dekker A G, Phinn S R and Brando V E 2014 A review of ocean color remote sensing methods and statistical techniques for the detection, mapping and analysis of phytoplankton blooms in coastal and open oceans *Prog. Oceanogr.* **123** 123–44

[4] Odermatt D, Gitelson A, Brando V E and Schaepman M 2012 Review of constituent retrieval in optically deep and complex waters from satellite imagery *Remote Sens. Environ.* **118** 116–26

[5] Malenovský Z, Rott H, Cihlar J, Schaepman M E, García-Santos G, Fernandes R and Berger M 2012 Sentinels for science: Potential of Sentinel-1, -2, and -3 missions for scientific observations of ocean, cryosphere, and land *Remote Sens. Environ.* **120** 91–101

[6] Transon J, D'Andrimont R, Maugnard A and Defourny P 2018 Survey of Hyperspectral Earth Observation Applications from Space in the Sentinel-2 Context *Remote Sens.* **10** 157

[7] Giardino C, Brando V E, Gege P, Pinnel N, Hochberg E, Knaeps E, Reusen I, Doerffer R, Bresciani M, Braga F, Foerster S, Champollion N and Dekker A 2019 Imaging Spectrometry of Inland and Coastal Waters: State of the Art, Achievements and Perspectives *Surv. Geophys.* **40** 401–29

[8] Lavrova O Y, Mityagina M I and Uvarov I A 2014 Detection and recognition of various water types in Black Sea coastal zone and in lakes of Crimea based on hyperspectral data analysis *Sovrem Probl Distantsionnogo Zondirovaniya Zemli Kosmosa* **11** 135–47

[9] Ghezehegn S G, Steef P, Hommersom A, Nils D R, Culcea O and Krommendijk B 2014 Hyperspectral remote sensing for estimating coastal water quality: case study on coast of Black Sea, Romania *Remote Sensing for Agriculture, Ecosystems, and Hydrology XVI* Remote Sensing for Agriculture, Ecosystems, and Hydrology XVI vol 9239 (SPIE) pp 340–8

[10] Cogliati S, Sarti F, Chiarantini L, Cosi M, Lorusso R, Lopinto E, Miglietta F, Genesio L, Guanter L, Damm A, Pérez-López S, Scheffler D, Tagliabue G, Panigada C, Rascher U, Dowling T P F, Giardino C and Colombo R 2021 The PRISMA imaging spectroscopy mission: overview and first performance analysis *Remote Sens. Environ.* **262** 112499

[11] Giardino C, Bresciani M, Braga F, Fabbretto A, Ghirardi N, Pepe M, Gianinetto M, Colombo R, Cogliati S, Ghebrehiwot S, Laanen M, Peters S, Schroeder T, Concha J A and Brando V E 2020 First Evaluation of PRISMA Level 1 Data for Water Applications *Sensors* **20** 4553

[12] Niroumand-Jadidi M, Bovolo F and Bruzzone L 2020 Water Quality Retrieval from PRISMA Hyperspectral Images: First Experience in a Turbid Lake and Comparison with Sentinel-2 *Remote Sens.* **12** 3984

[13] Anon Shagany-Alibei-Burnas Lakes System | Ramsar Sites Information Service; Information Sheet on Ramsar Wetlands, 5 July 1998, https://rsis.ramsar.org/ris/763

[14] Hersbach H, Bell B, Berrisford P, Hirahara S, Horányi A, Muñoz-Sabater J, Nicolas J, Peubey C, Radu R, Schepers D, Simmons A, Soci C, Abdalla S, Abellan X, Balsamo G, Bechtold P, Biavati G, Bidlot J, Bonavita M, Chiara G, Dahlgren P, Dee D, Diamantakis M, Dragani R, Flemming J, Forbes R, Fuentes M, Geer A, Haimberger L, Healy S, Hogan R J, Hólm E, Janisková M, Keeley S, Laloyaux P, Lopez P, Lupu C, Radnoti G, Rosnay P, Rozum I, Vamborg F, Villaume S and Thépaut J 2020 The ERA5 global reanalysis *Q. J. R. Meteorol. Soc.* **146** 1999–2049

[15] Dierssen H M 2019 Hyperspectral Measurements, Parameterizations, and Atmospheric Correction of Whitecaps and Foam from Visible to Shortwave Infrared for Ocean Color Remote Sensing *Front. Earth Sci.* **7** 14

AUTHOR INDEX

Alexandrova, Stella 100
Buchvarova, M 29
Buchvarova, Marusya 14
Deliyski, V 8
Draganov, D 29
Fiziev, Plamen 46
Gancheva, Irina 108
Georgiev, Georgi 58
Gyulchev, G 8
Gyulchev, Galin 34
Ivanov, S 70
Ivanov, Simeon 75
Ivanov, Svetoslav 75
Kilifarska, N A 79, 88
Mishev, Stoyan 42
Nedkova, P 8
Peqini, K 88
Petrov, N 1
Simeonov, Radoslav 64
Toneva, Zh 70
Tsvetkov, Ts 1
Velichkova, Ts P 79
Yazadjiev, S 8